T0143310

Computer Systems for Healthcare and Medicine

RIVER PUBLISHERS SERIES IN INFORMATION SCIENCE AND TECHNOLOGY

Series Editors

K. C. CHEN
National Taiwan University
Taipei, Taiwan

SANDEEP SHUKLA
Virginia Tech
USA

CHRISTOPHE BOBDA
University of Arkansas
USA

Indexing: All books published in this series are submitted to Thomson Reuters Book Citation Index (BkCI), CrossRef and to Google Scholar.

The "River Publishers Series in Information Science and Technology" covers research which ushers the 21st Century into an Internet and multimedia era. Multimedia means the theory and application of filtering, coding, estimating, analyzing, detecting and recognizing, synthesizing, classifying, recording, and reproducing signals by digital and/or analog devices or techniques, while the scope of "signal" includes audio, video, speech, image, musical, multimedia, data/content, geophysical, sonar/radar, bio/medical, sensation, etc. Networking suggests transportation of such multimedia contents among nodes in communication and/or computer networks, to facilitate the ultimate Internet.

Theory, technologies, protocols and standards, applications/services, practice and implementation of wired/wireless networking are all within the scope of this series. Based on network and communication science, we further extend the scope for 21st Century life through the knowledge in robotics, machine learning, embedded systems, cognitive science, pattern recognition, quantum/biological/molecular computation and information processing, biology, ecology, social science and economics, user behaviors and interface, and applications to health and society advance.

Books published in the series include research monographs, edited volumes, handbooks and textbooks. The books provide professionals, researchers, educators, and advanced students in the field with an invaluable insight into the latest research and developments.

Topics covered in the series include, but are by no means restricted to the following:

- Communication/Computer Networking Technologies and Applications
- Queuing Theory
- Optimization
- Operation Research
- Stochastic Processes
- Information Theory
- Multimedia/Speech/Video Processing
- Computation and Information Processing
- Machine Intelligence
- Cognitive Science and Brian Science
- Embedded Systems
- Computer Architectures
- Reconfigurable Computing
- Cyber Security

For a list of other books in this series, visit www.riverpublishers.com

Computer Systems for Healthcare and Medicine

Editors

Piotr Bilski

Warsaw University of Technology
Poland

Francesca Guerriero

University of Calabria
Italy

River Publishers

Published, sold and distributed by:
River Publishers
Alsbjergvej 10
9260 Gistrup
Denmark

River Publishers
Lange Geer 44
2611 PW Delft
The Netherlands

Tel.: +45369953197
www.riverpublishers.com

ISBN: 978-87-93519-31-2 (Hardback)
    978-87-93519-30-5 (Ebook)

©2017 River Publishers

Contents

**6 Gabor-Filter-based Longitudinal Strain Estimation
from Tagged MRI**   **129**

Łukasz Błaszczyk, Konrad Werys, Agata Kubik
and Piotr Bogorodzki

**7 A Decision Support System for Localisation and Inventory
Management in Healthcare**   **141**

Francesca Guerriero, Giovanna Miglionico
and Filomena Olivito

**8 Deep Learning Classifier for Fall Detection Based
on IR Distance Sensor Data**   **169**

Stanisław Jankowski, Zbigniew Szymański, Uladzimir Dziomin,
Paweł Mazurek and Jakub Wagner

Preface

Human healthcare and well-being are currently primary concern of governments and non-governmental organizations in the developed countries. Ageing societies require much more attention of medicine and therapy specialists than before. The prognosis for upcoming decades leaves no doubt that the assistance and permanent help for the elderly will refer to the constantly growing number of citizens. This calls for the systematic solutions that might alleviate the forthcoming crisis.

Computer technologies have been rapidly growing during the past decades, achieving the high level of applicability. Ubiquitous embedded solutions help people in multiple everyday tasks and activities. This allows for using them in any data acquisition and processing system, also in the field of healthcare. Required components of such a module include not only the computer part, but also sophisticated sensor system and the software algorithms for processing incoming data. The implementation of the complete system for the monitoring of the elderly or disabled person poses many new challenges: accuracy (allowing for raising alarms when the person requires medical assistance), discretion (hiding personal details of patients), or computational complexity (enabling the on-line mode of operation). These problems are currently solved suing various combinations of techniques and tools.

The following monograph presents the state of the art and advancement in the application of computer systems to support the healthcare and medicine systems, cooperating with the growing number of patients. The main parts of such a system (i.e. data acquisition, computing hardware and algorithms) are covered subsequently in ten chapters. The book starts with the general view on the current status of technology with the focus on the Ultra-Wide Band (UWB) equipment (chapter "UWB-Radar Monitoring of Movements in Homes of Elderly and Disabled People: A Health Care Perspective" by Tobba T. Sudmann et al.), where the techniques of tracking moving persons discreetly and accurately are presented. Next, the generic architecture of the elderly persons monitoring system using the wireless data transmission technologies is presented (chapter "System for Elderly Persons Behaviour

Wireless Monitoring" by Jerzy Kołakowski et al). The colour analysis and processing for the monitoring the condition of the living organisms is described by Alexey Lagunov et al. (chapter "Polychromatic LED Device for Measuring the Critical Flicker Fusion Frequency"). The next two chapters cover detailed implementation of the UWB technology for the position and physical condition of the monitored persons. The former topic is presented by Jerzy Kołakowski et al. (chapter "EIGER Indoor UWB-Positioning System"), while the latter – by Jan Szczyrek and Wiesław Winiecki (chapter "On Detection and Estimation of Breath Parameters Using Ultrawide Band Radar"). The data acquisition perspective on the monitoring of the elderly and disabled people is concluded with the implementation of the advanced signal processing techniques to process results of MRI.

The second part of the book covers details of the software part of the system, mainly consisting in the intelligent data processing. First, the generic architecture of the expert system supporting localization of the inventory in healthcare is presented (chapter "Decision Support System for Localisation and Inventory Management in Healthcare" by Francesca Guerriero et al.). Next, the deep learning artificial neural networks to detect falls of the elderly people is considered (chapter "Deep Learning Classifier for Fall Detection Based on IR Distance Sensor Data" by Stanisław Jankowski et al.). The similar problem solved by the decision trees classifier is discussed by Piotr Bilski et al. (chapter "Decision Trees Implementation in Monitoring of Elderly Persons Based on the Depth Sensors Data"). Finally, the application of artificial neural networks to control the prototherapy process is presented by Alexander Trunov (chapter "Recurrent Approximation in the Tasks of the Neural Network Synthesis for the Control of Process of Phototherapy").

The works presented in the book belong to applied sciences, in most cases considering working systems, implemented with the help of sophisticated hardware and software solutions. They show the current level of the healthcare support by the computer systems and provide insights for its future development. The book should be the help for both theoreticians and practitioners in the field of computer technologies' applications in healthcare and medicine.

Editors
Piotr Bilski
Francesca Guerriero

Acknowledgements

I wish to thank my mentors and guides in the world of science and engineering, for their constant help and inspiration: prof. Wiesław Winiecki and prof. Jacek Wojciechowski (who sadly passed away two years ago). Their expertise and kindness were an indispensable help in my all professional efforts.

Also, I would like to thank my whole family, especially my wife, Anna, whose inhuman patience and endurance allowed me to realize the things I thought were impossible. The same goes to my parents, brother Adrian and the rest of the family, who always stood for me.

Finally, I express my deepest gratitude to In Flames, Dark Tranquillity, Disarmonia Mundi, The Duskfall, Before the Dawn, Nightrage, Norther, and many other Melodic Death Metal bands. The ability to listen to their music is one of the reasons to live for.

Piotr Bilski

The authors would like to thank the anonymous referees for their valuable comments. The authors are grateful to project TeSS (Tecnologie a Supporto della Sanità – PON04a3 00424) for financial support to the work. The authors are also grateful to the ASP of Cosenza (Italy) for providing the samples.

Francesca Guerriero

List of Contributors

Agata Kubik, *1) Institute of Radioelectronics and Multimedia Technology, Warsaw University of Technology, Warsaw, Poland*
2) The Cardinal Stefan Wyszyński Institute of Cardiology, Warsaw, Poland

Aleksander Volkov, *Northern (Arctic) Federal University named after M. V. Lomonosov, 163002 Severnaya Dvina Emb. 17, Arkhangelsk, Russia*

Alexander Trunov, *Medical Equipment and System Dept., Petro Mohyla Black Sea National University, Mykolayiv, Ukraine*

Alexey Lagunov, *Northern (Arctic) Federal University named after M. V. Lomonosov, 163002 Severnaya Dvina Emb. 17, Arkhangelsk, Russia*

Angelo Consoli, *Eclexys SAGL, Via dell'Inglese 6, 6826 Riva San Vitale, Switzerland*

Dmitry Fedin, *Northern (Arctic) Federal University named after M. V. Lomonosov, 163002 Severnaya Dvina Emb. 17, Arkhangelsk, Russia*

Filomena Olivito, *Dipartimento di Ingegneria Meccanica Energetica e Gestionale, Università della Calabria, Italy*

Francesca Guerriero, *Dipartimento di Ingegneria Meccanica Energetica e Gestionale, Università della Calabria, Italy*

Francesco Piazza, *Saphyrion SAGL, Strada Regina 16, 6934 Bioggio, Switzerland*

Frode F. Jacobsen, *1) Centre for Care Research – Western Norway, Western Norway University of Applied Sciences (HVL), N-5020 Bergen, Norway*
2) VID Specialized University, N-0319 Oslo, Norway

Giovanna Miglionico, *Dipartimento di Ingegneria Informatica, Modellistica, Elettronica e Sistemistica, Università della Calabria, Italy*

Ingebjørg T. Børsheim, *1) Centre for Care Research – Western Norway, Western Norway University of Applied Sciences (HVL), N-5020 Bergen, Norway*
2) Department of Occupational Therapy, Physical Therapy and Radiography, Western Norway University of Applied Sciences, N-5020 Bergen, Norway

Jakub Wagner, *Institute of Radioelectronics and Multimedia Technologies, Faculty of Electronics and Information Technology, Warsaw University of Technology, ul. Nowowiejska 15/19, 00-665 Warsaw, Poland*

Jan Jakub Szczyrek, *Institute of Radioelectronics, Faculty of Electronics and Information Technology, Warsaw University of Technology, Warsaw, Poland*

Jaouhar Ayadi, *Eclexys SAGL, Via dell'Inglese 6, 6826 Riva San Vitale, Switzerland*

Jerzy Kołakowski, *Institute of Radioelectronics, Nowowiejska 15/19, 00-665 Warsaw, Poland*

Karol Radecki, *Institute of Radioelectronics, Nowowiejska 15/19; 00-665 Warsaw, Poland*

Knut Øvsthus, *1) Centre for Care Research – Western Norway, Western Norway University of Applied Sciences (HVL), N-5020 Bergen, Norway*
2) Department of Electrical Engineering, Western Norway University of Applied Sciences, N-5020 Bergen, Norway

Konrad Werys, *1) Institute of Radioelectronics and Multimedia Technology, Warsaw University of Technology, Warsaw, Poland*
2) The Cardinal Stefan Wyszyński Institute of Cardiology, Warsaw, Poland

Lorenzo Moriggia, *Saphyrion SAGL, Strada Regina 16, 6934 Bioggio, Switzerland*

Ludmila Morozova, *Northern (Arctic) Federal University named after M. V. Lomonosov, 163002 Severnaya Dvina Emb. 17, Arkhangelsk, Russia*

Łukasz Błaszczyk, *1) Faculty of Mathematics and Information Science, Warsaw University of Technology, Warsaw, Poland*
2) Institute of Radioelectronics and Multimedia Technology, Warsaw University of Technology, Warsaw, Poland

Lukasz Malicki, *Knowledge Society Association, Grazyny 13/15 lok. 221, 02-548 Warsaw, Poland*

Magdalena Berezowska, *Institute of Radioelectronics, Nowowiejska 15/19; 00-665 Warsaw, Poland*

Nadejda Podorojnyak, *Northern (Arctic) Federal University named after M. V. Lomonosov, 163002 Severnaya Dvina Emb. 17, Arkhangelsk, Russia*

Paweł Mazurek, *Institute of Radioelectronics and Multimedia Technologies, Faculty of Electronics and Information Technology, Warsaw University of Technology, ul. Nowowiejska 15/19, 00-665 Warsaw, Poland*

Piotr Bilski, *Institute of Radioelectronics and Multimedia Technologies, Warsaw University of Technology, ul. Nowowiejska 15/19, 00-665 Warszawa, Poland*

Piotr Bogorodzki, *Institute of Radioelectronics and Multimedia Technology, Warsaw University of Technology, Warsaw, Poland*

Ryszard Michnowski, *Institute of Radioelectronics, Nowowiejska 15/19; 00-665 Warsaw, Poland*

Stanisław Jankowski, *Warsaw University of Technology, Faculty of Electronics and Information Technology, Nowowiejska 15/19, 00-665 Warszawa, Poland*

Tobba T. Sudmann, *1) Centre for Care Research – Western Norway, Western Norway University of Applied Sciences (HVL), N-5020 Bergen, Norway*
2) Department of Social Science and Social Education, Western Norway University of Applied Sciences, N-5020 Bergen, Norway

Tomasz Ciamulski, *1) Centre for Care Research – Western Norway, Western Norway University of Applied Sciences (HVL), N-5020 Bergen, Norway*
2) Department of Electrical Engineering, Western Norway University of Applied Sciences, N-5020 Bergen, Norway

Uladzimir Dziomin, *Brest State Technical University, Department of Intelligent Information Technology, Moskovskaja str. 267, 224017 Brest, Belarus*

Vitomir Djaja-Josko, *Institute of Radioelectronics, Nowowiejska 15/19, 00-665 Warsaw, Poland*

Vladimir Terehin, *Northern (Arctic) Federal University named after M. V. Lomonosov, 163002 Severnaya Dvina Emb. 17, Arkhangelsk, Russia*

Wiesław Winiecki, *Institute of Radioelectronics and Multimedia Technologies, Faculty of Electronics and Information Technology, Warsaw University of Technology, ul. Nowowiejska 15/19, 00-665 Warsaw, Poland*

Zbigniew Szymański, *Warsaw University of Technology, Faculty of Electronics and Information Technology, Nowowiejska 15/19, 00-665 Warszawa, Poland*

List of Figures

List of Tables

List of Abbreviations

ADL	Activities of Daily Living
AI	Artificial Intelligence
ANN	Artificial Neural Networks
ANS	Autonomic Nervous System
ASP	Azienda Sanitaria Provinciale
B.C.	Before the birth of Christ
BoS	Base-of-Support
CFFF	Critical Flicker Fusion Frequency
CNS	Central Nervous System
CoM	(The body's) Centre-of-Mass
CT	Control Terminal
CU	Control Unit
CV	Cross Validation
DAQ	Data Acquisition
DCM	Dilated Cardiomyopathy
DSS	Decision Support System
DT	Decision Tree
EEA	European Economic Area
EEPROM	Electrically Erasable Programmable Read-Only Memory
EMW	Electromagnetic Wave
FL	Fuzzy Logic
FN	False Negative
FOQ	Fixed Order Quantity
FP	False Positive
FP	Fixed Period
HCM	Hypertrophic Cardiomyopathy
HMM	Hidden Markov Model
HSE	Health, Safety, and Environment
HVL	Western Norway University of Applied Sciences
IIP	Integrated Inventory Problem
IR	Infra-Red

LCD	Liquid Crystal Display
LED	Light-Emitting Diode
LGE	Late Gadolinium Enhancement
LISA	Linearly increasing startup angles
LOOCV	Leave One Out Cross Validation
LS-SVM	Least Square Support Vector Machine
LVOT	Left Ventricular Outflow Track
MATLAB	Matrix laboratory, a multi-paradigm numerical computing environment and fourth-generation programming language
MCIP	Multi-product Capacitated Inventory Problem
MLP	Multi-Layered Perceptron
MR	Magnetic Resonance
MRI	Magnetic Resonance Imaging
NArFU	Northern (Arctic) Federal University named after M.V. Lomonosov
NBC	Naïve Bayes Classifier
NN	Neural Network
NPCA	Non linear Principal Component Analysis
PC	Personal Computer
PCA	Principal Component Analysis
PIR	Passive Infra-Red
R&D	Research and Development
RAD	Dedicated hardware platform
RANN	Recurrent Artificial Neural Network
RF	Radio Frequency
RGB	Red, Green, Blue
RRSSCV	Repeated Random Sub Sampling Cross Validation
RVM	Relevance Vector Machine
SIFT	Scale-Invariant Feature Transform
SNS	Sympathetic division of ANS
SPAMM	Spatial modulation of magnetization
SSDM	System Support Decision Making
SSFP	Steady State Free Precession
SVM	Support Vector Machines
SWC	Synaptic Weight Coefficients
tMRI	tagged Magnetic Resonance Imaging
TN	True Negative
TP	True Positive

UNIX	A family of multitasking, multiuser computer operating systems
USART	Universal Synchronous-Asynchronous Receiver/Transmitter
USB	Universal Serial Bus
UWB	Ultrawide Band
WHO	World Health Organization
WUT	Warsaw University of Technology

Ultra-Wide Band Radar Monitoring of Movements in Homes of Elderly and Disabled People: A Health Care Perspective

Tobba T. Sudmann[1,2], Ingebjørg T. Børsheim[1,3], Tomasz Ciamulski[1,4], Jakub Wagner[5], Knut Øvsthus[1,4] and Frode F. Jacobsen[1,6]

[1]Centre for Care Research – Western Norway, Western Norway University of Applied Sciences (HVL), N-5020 Bergen, Norway
[2]Department of Social Science and Social Education, Western Norway University of Applied Sciences, N-5020 Bergen, Norway
[3]Department of Occupational Therapy, Physical Therapy and Radiography, Western Norway University of Applied Sciences, N-5020 Bergen, Norway
[4]Department of Electrical Engineering, Western Norway University of Applied Sciences, N-5020 Bergen, Norway
[5]Institute of Radioelectronics and Multimedia Technologies, Faculty of Electronics and Information Technology, Warsaw University of Technology, ul. Nowowiejska 15/19, 00-665 Warsaw, Poland
[6]VID Specialized University, N-0319 Oslo, Norway

Abstract

The task of this interdisciplinary Polish-Norwegian project is to develop radar technology for care services, and relates to telecare as a part of ambient-assisted living. The project has mainly an exploratory design, where the capabilities of the radar technology are tested out, related to elderly or disabled people, identified as in present or future need of assistive care. A system for preventing and detecting falls, or detecting other potential injurious situations, e.g. long lies, is its main goal.

Globally, health care policies suggest strengthening community-based services and home-based care. Ageing at home and health promoting lifestyle adds to longevity and compression of morbidity. Facilitating innovation and appropriation of ambient technology is part of these policies, which is expected to add value to uptake of new healthy habits and to early detection of functional decline. Ambient-assisted living may include systems for activity reminders, activity registration, health behaviour support (e.g. taking medication), fall prevention, detection of gradual functional loss and increased fall risk, and rising of alarms. The Radcare technology has the capability to produce data to meet these demands. This chapter will convey some experiences regarding opportunities and challenges in this interdisciplinary project.[1]

Keywords: Care services, Elderly and disabled people, Ambient-assisted living, Movement analysis, Radar technology.

1.1 The Relevance of Radar Technology and other Assistive Technology for Elderly and Disabled People

The interdisciplinary Radcare project has examined new possibilities for employing a sensor system based on impulse radar for care services [1, 2]. From the outset, the target fields were preventive care and detection of significant changes in the observed person's health condition, e.g., rapid physical decline, remaining in horizontal position after falls or seizures. Impulse radar technology are designed to observe human beings without using cameras or body worn sensors, and to measure human body movements and selected bodily functions from a distance, i.e., without attaching sensors or touching the body of the person concerned. The Radcare technology is expected to facilitate new approaches to care and assisted living, and, henceforth, add value to innovations in health and social care services.

The basic research and development of the Radcare technology; hardware, software, algorithms, and feasibility, have been executed by the Polish team at Warsaw University of Technology (WUT), of whom all are within the field of radio-electronic engineering. Relevant information and research-based knowledge on the assumed needs of the persons intended to use the Radcare

---

[1]This work has been supported by EEA Grants – Norway Grants financing the project PL12-0001 (http://eeagrants.org/project-portal/project/PL12-0001). The involved institutions are the Warsaw Technological University (WUT), with Prof. Wieslaw Wieniecki as principal investigator and The Center of Care Research – Western Norway at Bergen University College, with Prof. Knut Øvsthus as principal investigator, Associate Prof. Dr. Tobba Sudmann, Assistant Prof. Ingebjørg Børsheim and Associate Prof. Tomasz Ciamulski as co-investigators, and Prof. Frode F. Jacobsen as project leader.

technology in the future was delivered by the Norwegian health care team at Western Norway University of Applied Sciences (HVL), comprised of an occupational therapist (IBT), a physical therapist/sociologist (TTS), and a nurse/anthropologist (FFJ). The cooperation between WUT and HVL started from basic research on innovative applications of this new technology in the health-care area.

The need for the research and development on new technology in care services for elderly and disabled people relates to the ageing of the European population. We are presently viewing an increased need for complex care services due to the longevity of the population, and a rising number of elderly who will suffer from several chronic conditions, disability or frailty. Today, we observe a marked tendency towards compression of morbidity, i.e., more people live longer and an increasing number of them are disabled or frail only a shorter time at the end of their lives (from a few weeks to a few years [3]). While the European population aged 65+ is expected to rise, there is a decline in the age group 0–14. Statistics from the European Union suggests that by 2060 one in eight persons will be above 80 years, and nearly one in three persons will be above 65 years [4, 5]. This statistic does not include Poland and Norway, or the other EEA countries. To our knowledge, the same trends are expected also in the EEA countries. An ageing of the population is a positive and unique result of a rising standard of living. However, this creates challenges pertaining to care of people at the end of their life, and to offer care and health service to any person in need of assisted everyday living. We might face a mismatch between the number of persons in need of assistance, and the number of people capable of assisting, given that yesterday's way of providing and organising care services pertain into the coming decades.

Falls are one of the main causes of hospitalisation of elderly people [6–14]. The number of deaths globally caused by fall events was around 391.000 in 2003, where approximately 40% of the falls were from people over 70 years of age [15, 16]. In Norway, a new Report to the Storting (The Parliament) suggests that it is possible to reduce number of falls with 40% by appropriating diverse preventive measures [17], including technology. Local governments across Europe and the World Health Organisation (WHO) have launched a series of policy frameworks that recognises and recommends that treatment and professional care should be transferred from specialised care services to municipal health and care services. This transference will secure that patients are offered health and care services that are better suited to their personal needs and to the local context. Additionally, a welcomed increase in use of technology and innovations in care services open for new possibilities for designing content and provision of services, including a participatory design, i.e., end-users are active partners in the design process. The normative platform for the Radcare study is to make care smarter and personalised. The personally adjusted health care service offers new opportunities for health promotion, health behaviour change, and a lifelong uptake of new knowledge and lifestyle by the receivers [18, 19]. Appropriation of health technology may reduce the pressure on health and care services, as shown in the Oslo pilot study [20].

Recently published Norwegian policy documents suggest that future health and care needs should be met by strengthening health promotion and community-based services, by acknowledging next-of-kin and informal home-based care [21, 22]. These aims are supposed to be reached by inter-alia facilitating innovation and appropriation of ambient technology. The Radcare technology answers to some of these challenges.

Health, social, and welfare services are expected to promote public health and enhance people's abilities to lead their lives as they prefer, sustaining healthy everyday living and exercise citizenship throughout their life span. People with disabilities have asserted growing up and ageing in place, independent living, and social participation as rights for several decades. Contemporary disability scholars justify their claims by providing evidence for the interrelationship between persons, places, social participation, and wellbeing [23]. Suggestions in national policy documents are to comply with what has been asserted by disability organisations. Globally increasing demands on welfare services have purported several political initiatives from the WHO and national governments. Reorganisation of health, allocation of resources towards health promotion, reviving community-based care and encouraging person-tailored health services, are implemented to improve public health and reduce costs [22, 24].

Health-in-all policies are implemented across the public sector, as a turn in public health promotion is observed, encouraging the public to acknowledge responsibility for their own present and future health status. WHO's policy is active ageing, a process of optimising opportunities for participation, health, and security to enhance quality of life, emphasising the community as a key setting for interventions [7]. Active ageing shifts strategic planning away from the "needs-based" approach (which assumes that older people are passive targets) to a "rights-based" approach that recognises the rights of people to equality of opportunity and treatment in all aspects of life, as they grow older. It supports their responsibility to exercise participation in the political process or other aspects of community life [24, 25]. The Radcare technology is promising as a support in active ageing, ageing at home and to uptake and sustainment of healthy behaviour.

1.2 Healthy Ageing: Ageing at Home

The Radcare project acknowledges that ambient (assistive) technology in homes may support end-users' ability to live at home. This technology is already part of homes, workplaces, or car environment; e.g., fire or burglar alarms, movement detectors, door or window sensors, temperature or humidity sensors, movement sensitive lighting, adjustable heights and lengths, Internet of Things, robot technology, intelligent chairs, navigation, smart phones, computers, gesture controlled systems, etc. Most of these technologies are installed and appropriated for security, energy saving, reduced workload, remote control of individual preferences, and communication or multimedia purposes. When implementing these technologies people usually do not pay attention to the compatibility or communication between sensors. Usually, there is no need to consider communication between local devices and distant receivers or other contextual factors. However, some of these technologies are installed and appropriated for around-the-clock assistance to secure or promote independent living [18, 20, 24, 26–30].

An increasing part of the population will benefit from care and health-related technology during their lifespan, or for shorter, intermitted periods. Part of this technology will be provided for or commissioned by municipalities, private-for-profit, private-non-profit, or non-governmental organisations, or obtained by persons concerned about their families. As Van Hoof et al. [31] state, ambient technologies contribute to increased safety and security at home.

Policy frameworks around the world put increasing emphasis on healthy ageing, and are stressing the responsibility everyone, family, or community must take on healthy habits to keep themselves active and healthy. The key pillars from the WHO policy framework – participation, activity, and safety – are believed to reduce the prevalence of non-communicable diseases, frailty, and obesity. Participation and activity are the keys to successful reablement, as to will and ability to keep up new habits or treatment regimens. Activity diary and registration seem to motivate most people. Living is moving, and every movement imply temporarily loss and regaining of balance. I.e. falls due to inefficient balance reactions is to be expected if people loose physical function and/or are not careful enough, or are having bad luck. Healthy ageing entails an increasing amount of activities in common rooms, outdoors, and in the community. The flipside of ageing at home is that injurious falls must be expected. The Radcare technology may serve to prevent falls, and to reduce the risk of increased morbidity or mortality after a fall by rising alarms and preventing long lies.

Recommendations for persons 65+ are light to moderate physical activity [31–33]. It is the only measure known to prevent, slow down, and ameliorate cognitive decline, and efficient when it comes to prevention of falls [18, 34]. Physical activity is a hallmark of healthy ageing and successful ageing at home, and the Radcare technology can be used for these purposes as well.

1.3 Definition of Falls and Movement Analysis

Generic movement analysis with Radcare needs to consider the epidemiology of falls [35, 36]. People fall because of age, frailty, reduced range of motion in joints, muscles atrophy, slower balance, or compensative reactions. Some of these functional losses are due to under-usage of the muscle skeleton apparatus, often preceded or followed by morbidity. Reduced mobility in the trunk and major joints reduces flexibility and speed in regaining balance, and increases the risk of falling. Intrinsic causes (biological) are often at work together with extrinsic causes (material, behavioural, environmental, and socio-economic). According to Rubenstein [9], more than half of all fall risks are modifiable; falls due to accidents or environment (31%), due to gait or balance disturbance (17%), or dizziness (13%). Reducing the numbers of falls decreases overall morbidity, pain, mortality, and costs.

Igual et al. [37] point to several challenges, issues, and trends in fall detection systems. Amongst them is the call for awareness of the limitations and possibilities of the systems, e.g. user acceptability and performance under real-life conditions. Skubic et al. [38] highlight the need for creating robust systems for translating from physical measure to clinical space, and the challenges arising from different aims in monitoring, e.g., physical assessment, mental assessment, cognitive health assessment, or management of chronic conditions. Even though these might overlap, the need for different sensors or data processing methods ought to be considered. The mentioned research groups acknowledge the need for the interdisciplinary research.

A commonly agreed definition of a fall mechanism is an incorrect alignment of centre-of-mass and base-of-support. A fall may also be interpreted as an unwanted movement; i.e., movement analysis is warranted to understand how a fall evolves and how it can be intercepted or detected. The FARSEEING consensus [39, 40] states that a fall is an unexpected event in which the person comes to rest on the ground floor or lower level. A five-phase model is developed accordingly (Figure 1.1).

A fall has a different trajectory whether its direction is forward, sideways, backward, or different combination of directions. Regaining balance sideways

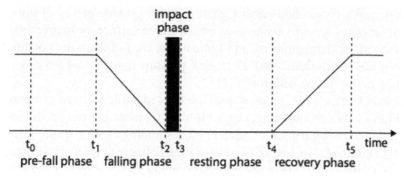

Figure 1.1   Fall event model [40].

increases the risk of falling, i.e., a fall trajectory in a different direction than the intended movement. Most falls have unpredictable trajectories, because the faller will try to hinder or dampen the event. The Radcare might add value to health care in all phases: as a pre-fall prevention (pre-fall phase), as a feed-forward warning to the person (falling phase), by triggering alarms to a call centre (resting phase), or in the recovery phase (post-fall prevention). The shorter the lie or resting phase, the less the risk for increased morbidity and mortality after a fall.

Falls can also be classified as injurious or non-injurious. Most are non-injurious at impact, but may become injurious if fallers are unable to recover and rise from the position they acquired during the fall. From a health care perspective, there are two pressing issues: prevention of falls and reduction or elimination of long lies. A technology that correctly can detect the latter and raise alarms has the potential to save lives, and reduce post-fall morbidity. People fall due to multiple causes, whereof most are impossible to eliminate from the persons living environment or their body, as is exemplified below.

Falls happen when there is a change of movement direction, and/or a change in activity or attention. People fall more often in bedrooms and bathrooms. In bedrooms, there are demands on the body's biological capability for change in position towards gravity, i.e., regulation of heartbeat, blood pressure, and muscle tone to prevent the person from fainting or dizziness, and a demand for dynamic balance and coordinated movements and attention. On the other hand, bathrooms often have hard-tiled floors, heating cables beneath the flooring, furniture, and fittings made of hard materials. Hard surfaces can increase injuries, and warm floors can induce burn damage during a long lie. Bedrooms and bathrooms are usually small, and (in most homes) there is less room for compensatory movements, which might increase the risk for falling.

Conversely, small rooms and narrow passages also provide plenty of possibilities for support – directly by holding onto a stable surface, or indirectly by passive support, or dampening of a fall (changing the fall from the potential injurious to a non-injurious one). High-risk persons have a doubled risk of injury when falling in the bath room [11].

Robinovitch et al.'s [41] study of real life falls identified several common causes of falling indoor including, but not limited to, incorrect weight shifting (41%), walking (24%), trip or stumble (21%), standing (13%), sitting down (12%), hit or bump (11%), loss of support (11%), collapse (11%), and slipping (3%). As Rubenstein [9] correctly observed, many of these falls are due to modifiable risk factors.

Examples of fall-related injuries are hip fractures, which often is a sign of frailty, and where a long lie leads to a serious increase in morbidity and mortality. Fractures of the upper limbs or head injury lead to pain disability and loss of independence. This calls for the increased need for assistance, and henceforth, increased public and private expenditure. Rapid detection and treatment of injuries reduces morbidity and mortality.

Following Rubenstein [9], the epidemiology of falls is a call for inter-disciplinary collaboration in the assessment and interventions, particularly exercise, attention to co-existing medical conditions, environmental inspection, and hazard abatement. Fall risks can be decreased, particularly by the systematic low to moderate intensity of physical activity.

1.3.1 Activities of Daily Living and Falls

Activities of daily living (ADL) are conceived of as everything that people do, including looking after themselves (self-care), enjoying life (leisure), and contributing to the social and economic fabric of their communities (productivity). As outlined above, fall prevention and detection is cognate to the movement analysis. To personalise fall prevention or detection, the technology must recognise and appreciate all the bodily movements of everyday living.

Activities of daily living include everything one does, livelihood, living conditions, friends, families, and can be detailed by the personal or instrumental ADL. The former (pADL) relates to the maintenance of personal wellbeing; hygiene, eating, and dressing. These activities often take place in the bedroom or bathroom. The latter (iADL) relates to using props, utensils, shopping, driving, cooking, washing, using phones, or assistive technology. These activities may be found anywhere, including outdoor. A technology designed to detect unwanted events, or to motivate uptake and sustainment of healthy behaviour must discriminate between ADL and unwanted movements or events.

There is a positive effect of physical activity on all non-communicable diseases. It is particularly important to take action on the correlation the between physical activity and cognitive decline [42, 43]. Community dwelling is not to be homebound, and preventive technology must consider this. Furthermore, long-term care is not to stay in bed 24/7, and patients are expected to move around indoors and outdoors. Understanding of human movements means expecting movement variability, falls, and hence falls variability. Acquired movement data can be used for prevention, changing health behaviour, and fall detection.

Movement analysis aims at determining how the body's centre-of-mass (COM) and base-of-support (BOS) are aligned or misaligned during everyday living, and to appreciate the exposed variability. Movement is exercising motor control and dynamic balance depends on the feed-forward and feed-back mechanisms [44]. The former includes anticipatory postural adjustments and accompanying postural adjustments. They are due to preparatory aroused neuromuscular activity, before a movement takes place, e.g. thinking about drinking. Accompanying adjustments happen during movement, e.g. adjusting the grip on the glass or the trajectory towards the mouth, adjusting speed, or rhythm of any movement. Prolonged or delayed accompanying postural adjustments are found for persons with fear of falling [45]. Anticipatory and accompanied postural adjustments and neuromuscular arousal is mod-ifiable through exercise and personalised "tricks" and reminders. Telecare and assistive technology can be appropriated as means for these ends.

Normal balance reactions include the following five strategies, which may become less efficient due to physical decline, under usage or environmental factors. Balance reactions must be appropriate to the missing alignment between BOS and COM. A too slow, moderate, or an exaggerated reaction, may increase the fall risks. Balance reactions have increasing complexity, from shifting weight or moving the ankles, bending, or stretching knees, stepping in any direction, grasping, and lastly, extending arms in protection or for dampening of the fall. These reactions are partly modifiable by physical activity.

1.3.2 Self-Selected Walking Speed: A Vital Sign

A vital sign tells if the observed person is alive, and is a quick assessment of vital functions [46–51]. Heartbeat, respiration frequency, temperature, and

Table 1. Normal gait speeds for healthy community-dwelling men and women.[11]

Age (years)	Gender	Average Gait Speed (m/s)
20-29	Men	1.36
	Women	1.34
30-39	Men	1.43
	Women	1.34
40-49	Men	1.43
	Women	1.39
50-59	Men	1.43
	Women	1.31
60-69	Men	1.34
	Women	1.24
70-79	Men	1.26
	Women	1.13
80-89	Men	0.97
	Women	0.94

Figure 1.2   Nominal walking speed for persons of various ages.

Source: http://lermagazine.com/article/self-selected-gait-speed-a-critical-clinical-outcome

blood pressure are vital signs; they give precise and quick information about the overall health status of the person. Some argue that pain is also a vital sign.

Walking is very complicated and involves all bodily functions. To be able to walk at a normal speed, the overall health status must be good. If any of the vital signs show abnormal scores, it will be possible to observe this by a quick assessment of how the person is walking. Due to this fact, gait is also a vital sign. We collected data on walking in self-selected speed from the persons being monitored by Radcare and use them to identify increased or decreased risk for falling or changes in health status. Reference values are established for all vital signs, including walking (see Figure 1.2 on age related walking speed).

The normal self-selected walking speed of a healthy person is <5 s/5 m. The speed is calculated over a predetermined distance at pre-set times. The average speed is calculated and compared to pre-determined critical values, see figures. Calculation of walking length and time during 24 h also provides important health relevant information, and for this reason, we included all those measures in our Radcare experiments.

1.4 Step-by-Step Development of the Radcare Technology

Experiments related to development of the Radcare technology are focused on movement analysis, based on the understanding of a fall as an unexpected event where the person comes to rest on the ground floor or a lower level [52]. Henceforth, fall detection and fall prevention are part of the movement analysis, i.e., assessment of physical, mental, and cognitive health. The paragraphs below give a short overview of the consecutive steps in the R&D process.

1. Opal sensors and movement analysis

   The first-fall detection experiments were set up in collaboration with physiotherapists at Bergen University, Norway, and performed using Ambulatory Parkinson's Disease Monitoring (APDM) technology with Opal [53] sensors and wireless control, readout and synchronisation of data from them. The sensors contained accelerators for each of three axes. Additional equipment was helpful in processing and interpretation of data from accelerators (gyroscopic and magnetic). The sensors were attached to the body at different places: hips, shoulders, and head. Using ourselves as models, we performed falls experiments. Collected raw data were sent from HVL/Norway to WUT/Poland for further processing. Based on recorded accelerations, trajectories of movement were calculated at WUT and used as input to algorithms under development, before first data from Radcare radar sensor would be available.

2. Qualisys motion capture and movement analysis

   The second fall experiment was set up in collaboration with physiotherapists in the movement laboratory at HVL, using Qualisys [54] motion capture system based on a set of highly sensitive infrared cameras which can observe markers reflecting the infrared radiation. We explored falls based on typical ADL situations and known fall risk situations, like stumbling (e.g., related to doorsills and carpets) and sitting to standing movements (e.g., from the bed or toilet). As above, we used ourselves as models. This broad focus covers falls due to intrinsic (e.g., low blood pressure), extrinsic causes (e.g., walking sticks and insufficient light). The test person was marked with approx. 20 markers in frontal, sagittal, and transversal planes (bilateral on ankles, knees, hips, shoulders, elbows, wrists, forehead, ears, back head, and the back). We had seven motions capture cameras and an area of approx. 3×3 m covered with thick Airex mats. The results from these experiments were also forwarded to WUT/Poland, to aid the R&D on software.

3. Radcare and ADL movements

   The third experiment was set up at HVL/Norway, using the Radcare technology developed at WUT/Poland. Ten different movement scripts were chosen, e.g., sit to stand, stand to sit, walking, turning, lying down, rising from floor, and sitting activities as drinking from a cup, doing one's hair, and tying/untying shoe laces. These data show that the first versions of the Radcare technology where not able to identify arms and legs, or "rapid" movements. The technology was not able to capture differences in positions if they were performed at normal speed. Rapid turns were also a challenge. The third series of experiments indicated that it would be possible to detect movement in different positions (lying, sitting, and upright), distance covered during monitored space, quality of gait (rhythm, sway). The technology holds potential to detect nocturnal, unwanted events as seizures (excessive movements) or fainting (no movements to detect), but has not been tested in real-life sleeping situations yet. The results were useful for further development of Radcare.

4. Radcare and gait analysis

The fourth experiment used version with upgraded software and focused particularly on the movement detection – where, how, how long, and covered area and distance. The technology now proved capable of detecting direction, speed, distance, turns, and standing still. This was a very promising step. We identified how large area the UWB-radar technology could cover and still produce high-quality data, planned for the fifth phase. The data acquired from acceleration signals in these experiments can be used to infer changes in physical and mental health of the observed persons. Unintelligible start and stops, changes in direction or repeated activity might serve as an alert to cognitive health. The observed changes do not correspond directly to a diagnosis, but can be used to infer the need for further assessment of mental and physical health.

5. Living lab: A real home test

The fifth experiment was set up in collaboration with one elderly lady living alone. With her written permission, the Radcare technology is set up in her home, monitoring a hall/walkway she must cross whenever moving between different rooms in her apartment: living room, kitchen, bathroom, and the bedroom. The installation demanded the involvement one researcher and one "generator"/handyman (her son) for several hours. The installation appeared obtrusive in the environment and cables proved not well fitted to the apartment. However, both the lady in question and her son and daughter in law have accepted that the provisional installation is acceptable for research purposes, even though it is obtrusive and poses risks. Cable in the staircase increased the risk for falling, it also required using the neighbour's internet access point and more

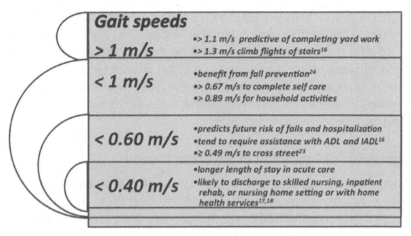

Figure 1.3 Relation between the walking speed and the risk of falling.

Source: http://lermagazine.com/article/self-selected-gait-speed-a-critical-clinical-outcome

electric plugs than recommended for one socket and increasing risk of fire. For comparative purpose, we assessed her self-selected walking speed with a stopwatch; in the hallway, in her kitchen and outdoors. The speed varied from 45 to 85 cm/s (Figure 1.3). She was easily motivated by our measurements, indicating that real-time feedback increased her speed and vigour. These are easy means for fall prevention and health promotion. The more vigorous lifestyle might increase the risk for falls. On the other hand, it increases possibilities for the quick recovery after a fall, and reduces the risk for injurious falls and long durations of lying down. A trade-off between falls risk, healthy ageing and ageing at home, justifies the possible increased risk. The woman in question and her family reached the same conclusion. As for the capability of the radar technology, the possibility for observing through not-metal walls is promising, potentially increasing the observational area considerably.

1.5 Discussion: Findings and Experiences

Observed changes in gait is a robust indicator of increased risk of falling and reduced physical, mental, or cognitive health [55–59]. Fear of falling are often correlated to the fear of moving [60, 61]. If an observed person moves very little, the assessment of fear of falling, moving, and gait quality is pertinent. This provides the information about all vital signs and gives a quick assessment of the cognitive health. Data provided by the Radcare technology can be used to infer some of the information, otherwise obtained by using standardised tests [62–65]. Information around the clock on the variability in movement (e.g., speed, distance, place, frequency, and turns) gives a quick overview of how the monitored person is spending their everyday lives compared to the previous data. The longer the observation time, the easier to make personalised reports and "red alerts" concerning health.

1.5.1 Detection of Presence at Selected Places

The two-dimensional location of monitored persons can be used to observe their presence in selected places in the household, such as the bedroom or the bathroom, at different times of the day. We can then determine a sequence of places identified in equal intervals of time, for example every 10 min, during the selected period, such as a single day. The comparison of such sequences obtained for consecutive days may allow for observing trends and anomalies, indicating an increase or decrease in health. These sequences can be presented to medical staff or analysed automatically using statistical methods [36].

Our system has proved capable of detecting people behind a non-metal wall. Although the accuracy of the location estimate is lower behind the wall than it is inside the room, in which radars were placed, it is sufficient for rough identification of the area. Therefore, the use of Radcare technology allows for measuring the presence at selected places in a household without the need to install sensors in each room. If radar sensors were to be combined with another source of data, such as Kinect cameras [66], this would also allow triggering the measurement by other devices only when the monitored person is within their range.

In the real life test situation we identified a pre-defined transfer zone [36, 67], the hallway, as an eminent place for observation. The inhabitant must cross the hallway to get to any room in the apartment (Figure 1.4). The red crosses above represent the two Radcare antennas' placements.

1.5.2 Detection of Motion

We assume that the patient is in motion, if the sum of distances between five consecutive location estimates is >20 cm. Our project demonstrated that the total amount of time, in which the patient is in motion during a selected period (for example, 24 h) can be measured and is a useful indicator of the patient's level of activity. Visualising its values for several consecutive days helps the health and care staff to observe an increase or decrease in the patient's health.

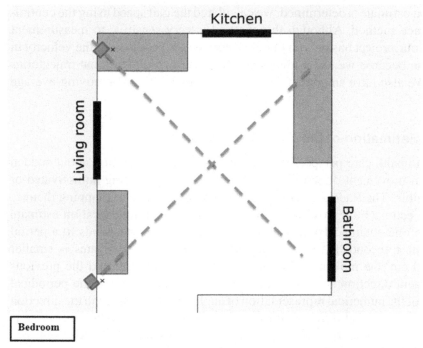

Figure 1.4 The test home had an ideal hallway for monitoring movement. The resident had to cross the sensors when ever leaving or entering a room or the front door (below the bedroom, outside the frame).

We measured the amount of time spent by the patient without any movement, allowing for detecting dangerous incidents, such as falls: an alarm can be activated if motion is not detected in the selected time interval, such as 1 h, except during the night. On the other hand, motion detection during the night can be used to assess the quality of sleep: many motion events detected during the night indicate bad sleep [36]. Poor sleeping quality is detrimental to a frail person, and increases risk for falling. The relationship between poor sleep and overall health cannot be inferred from radar data, and hence must be assessed by the health care personnel.

1.5.3 Estimation of Gait Speed

The gait speed and variability is an informative indicator of health [51, 55, 56, 68–74]. The use of radar technology allows for estimating the self-selected gait speed by observing the patient during everyday activities, in contrast to clinical tests, in which the person might perform differently. Every time a new location estimate is determined, we calculated the gait speed using the central-difference method. Although this method is very sensitive to measurement errors, our project has proven that it is sufficient for estimating the velocity in our case because we use it after smoothing the two-dimensional trajectories [75]. We also have smoothed the velocity estimates using a moving-average filter.

1.5.4 Estimation of Movement Direction

From a health care perspective, it is useful to get information about sudden shifts in movement direction, particularly the ones seeming unmotivated or intelligible. The Radcare technology has proved capable of mapping them.

We estimate the movement direction using the current location estimate and the one obtained 10 measurements earlier: this corresponds to a period of about 1 second. If the distance between these two estimates is smaller than 30 cm, the movement is considered as insignificant, and the previous movement direction is considered as unchanged [76]. Due to the periodical nature of the numerical representation of angles, a small change in the direction can cause a large difference in the representative number when the angle is close to π or $-\pi$. We used a procedure based on the MATLAB's *unwrap* function [77] to smooth such jumps.

Although the movement direction itself is not an informative quantity, it may provide data that can be used to infer vital information about the patient's ability to plan, remember, find and replace utensils, clothes or props, or indicate a tendency to deliria at specific times. Information on meals, medication, and sleeping habits can be used as a backdrop when interpreting movement data. Over- or under-medication can be documented, as well as dietary habits that have a negative impact on the health and wellbeing of the observed persons.

1.5.5 Estimation of Travelled Distance

The travelled distance can be calculated by summing up the distances between consecutive locations of the monitored person, which we did in the project. However, small deviations in the location estimates, caused by the imperfection of the measuring system, may lead to overestimated results. Therefore, instead of using raw values of the distances, we employed their projections on the current movement direction.

The estimates of the travelled distance can be summarised for a selected period, such as the last 8 h or the last week. The visualisation of the distance

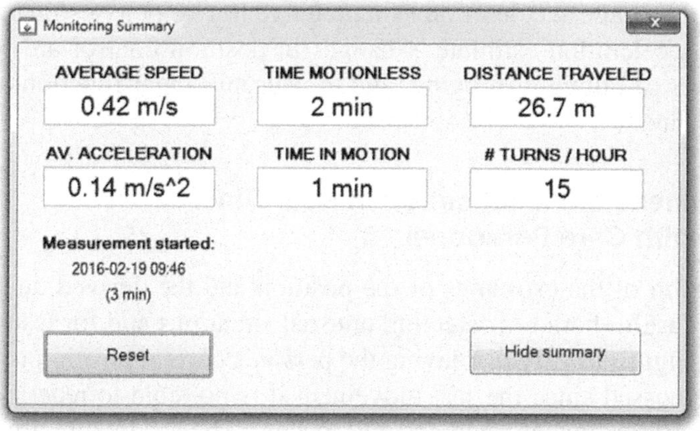

Figure 1.5   User interface for the person monitoring application.

travelled each day during a longer period may help the medical or healthcare staff detects a long-term trend indicating an improvement or deterioration of health.

Figure 1.5 illustrates how data obtained gives the person concerned, family, or health care personnel a quick overview of the amount and quality of movement within a specified timeframe. The latter may be circadian, per hour or other preferred measures that have a bearing on other significant events during the day and night, e.g., medication, meals, and sleep. In the example from the test home above, we can observe that the walking speed of the inhabitant is below a critical level. Ideally, a woman of her age should have a walking speed of 0.60–0.90 m/s. Manual registration revealed that she could keep a speed according to age, which makes it even more interesting to observe how she moves when not "tested" by the health care personnel.

1.5.6 Estimation of Acceleration

From a healthcare perspective, it could be interesting to get information about positive and negative acceleration. It is estimated in three types of situations: starting to walk, stopping, and changing the direction, can provide valuable information about the monitored person's balance, postural and motor control, and cognitive functioning. To estimate the acceleration, the WUT/HVL team differentiated the smoothed velocity estimates, $\hat{v}_n$, using the central-difference method, considering only the periods in which the monitored person starts or stops moving, or turns. Each time such an event was detected, we calculate a new value of average acceleration using the five last velocity estimates and the previous acceleration estimate. Changes in postural control are related to fall risk, fear of falling or moving, and to neuromuscular functioning and functional decline.

1.5.7 Usefulness of Visualisation in Real Time for Health Care Personnel

The visualisation of the estimates of the position and the derived quantities in real time is useful both for detecting unusual situations and for testing the software, see Figure 1.6. By displaying the person's current position together with the time passed since the last movement it is possible to react quickly when an unwarranted event, such as a fall, occurs. The acceptable time for a recovery phase is after a fall can be pre-set and an alarm can be triggered when movement data cease to be detected. Furthermore, by visualising estimates of the quantities such as movement speed and direction we can check their validity. Colours can be used for motivational purposes with the end user.

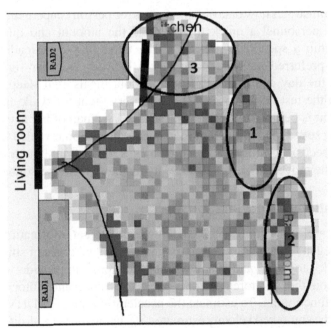

Figure 1.6   Visualisation of frequency of visiting locations in the apartment by the monitored person.

1.6 User Interface and Participatory Design

User acceptability and appropriation of ambient assistive technology must be considered [78–84] whenever offering telecare solutions to persons in their own homes or in long time care facilities. From the literature overview, we know that in the appropriation of telecare or other assistive technology, a series of considerations are done by the end users. A successful technology must honour non-technical expectations from them, of which a huge variability is to be expected. A few known facilitators and barriers include, but are not limited to, aesthetics, preferences, and vanity. The core technology should be as small as possible, and possible to fit into diverse boxing. The complete Radcare equipment should ideally be fitted into existing furniture and fitting in a seamless manner.

Ageing at home poses some ethical questions. Even if we want to use data around the clock to infer threats to the person's health, we will simultaneously collect data on intimate and personal habits. This must be explicitly vented with persons in question before the telecare is installed in their homes.

Telecare also poses challenges related to HSE, Health, Safety, and Environment. As the installation in the test home revealed, more cables and gadgets increase risk for falls. The Internet connection and capacity, numbers of available sockets and their placements also raise questions about HSE. Increased fall risks might be accompanied by the greater fire risk, or heating of sockets or plugs. Maintenance of the equipment is a known challenge, which must be considered before the installation of telecare in any home.

Uptake of telecare and healthy habits is motivated by personalised reports, activity reminders and diaries. Some persons will probably find it stimulating to add personal notes on activity, medication, or meals, whereas others are more than happy to receive simple visualisation, e.g., a traffic light. Telecare depends on the end users' acceptance and adherence, and must be designed to include some preferences. Barriers and facilitators are recognised and discussed continuously – with end users, health care personnel and telecare designers (hardware, software, and design). A participatory design holds the potential to increase motivation and documentation for uptake of healthy habits, shared decision-making, self-registration, planning ahead, and registering.

Physical activity is documented to modify risk factors for non-communicable diseases, cognitive decline, falls, and frailty [15, 31–33, 85]. However, uptake of lifestyle changes poses some challenges [18, 19, 86]. Telecare and interactive user interface are well designed for supporting motivation and uptake of new healthy habits.

1.7 Concluding Considerations and Suggestions for Future Research

Radar technology provides a promising path towards personalisation, well-being and ageing at home for elderly and disabled people in need of telecare assistance. Several promising areas for employment of radar technology have been identified, like movement analysis as basis for decision-making about preventive interventions before, under, or after the fall [59, 87]. Radar technology appears to be a promising candidate for personalised health monitoring, with simple communication to relevant stakeholders (e.g., friends, family, community service, general practitioner, and hospital).

The human–technology interface is complicated, and hence a multidisciplinary approach proved useful. End users' preferences and experiences must be taken into consideration continuously, e.g., by stating participatory design as a mandatory commitment. Input from the health sciences proved to be important for identification of significant movements and vital signs in daily life activities and for developing promising areas of observation and measurements. The most interesting results have derived from mapping the capability and feasibility of the Radcare technology to infer information that acts as "physical evidence" of the general health status and wellbeing. Measurements of daily life actions in the test home have proven to have clinical relevance, and to provide telecare support and assistance for living well and longer at home.

As suggested above, the Radcare technology can be appropriated as means for pre-fall prevention, contribute to rapid alarm and reduction of the long lie (fall injury prevention), the post-fall prevention, i.e. avoiding future falls by utilising acquired knowledge about the faller, and creating the personalised prevention and warning system. The possibility to calculate variance in walking patterns when monitoring a home-dwelling person is fruitful when participatory design is an option. Data collected can be used to design person specific pre-fall interventions and post-fall interventions [87].

Although the radar-based system yields less accuracy of the position estimates than the system based on the depth sensor, it allows for obtaining estimates of the healthcare-related quantities accurate enough to support the medical and healthcare staff in detecting long-term trends and anomalies.

Although the radar-based system yields less accuracy of the position estimates than the system based on the depth sensor, it allows for obtaining estimates of the healthcare-related quantities accurate enough to support the medical and healthcare staff in detecting long-term trends and anomalies.

However, radar sensors are usable only if the monitored person passes through the centre of the monitored area often enough, since the position estimates on the borders of this area are corrupted by large errors. Therefore, this solution can be used for estimating the presence at selected places away from the centre, as the accuracy is sufficient to roughly identify the area, including locations behind walls. A sensor fusion is also possible if the rough radar-based position estimates are used for enabling the acquisition of data by depth sensors only when the monitored person is within their range.

The Radcare technology can probably be used for several purposes not imagined when the project proposal was launched 3 years ago. A few suggestions for further use and research are listed below:

- Identification of health risks;
- Physical activity;
- Balance and strength;
- Fall risk assessment;
- Identifying risk behaviour;
- Pre-fall prevention intervention;
- Fall injury prevention ("red alert");
- Post-fall prevention intervention;
- Integration of data from radar and body-worn sensors;
- Activity registration – activity diary;
- Facilitation of health behaviour changes;
- Facilitation of adherence to preventive measures; and
- Participatory design – end users might suggest repurposed/different use for same or different ends.

The project aimed developing novel technology for the health and care sector needs to be designed as an interdisciplinary collaboration, where insight from health and social scientists and practitioners is as important as technical knowledge. Based on our experience from a project mandated to design a system for detecting falls by the analysis of everyday movements/ADL, we have illustrated some challenges, opportunities, and important lessons learned. The wider context of unwanted events or falls in the home needs to be taken into consideration while expanding the Radcare project; e.g. intrinsic, extrinsic and environmental factors, user-technology interface, personalisation of monitoring, and individual preferences and understandings of health, safety and wellbeing.

From our point of view, a scenario where Radcare technology and other types of technology in combination can contribute to information that can be used to prevent falls, is still an option, much needed [37, 38, 88].

Acknowledgements

This work has been accomplished within the project PL12-0001 financially supported by EEA Grants – Norway Grants (http://eeagrants.org/project-portal/project/PL12-0001).

References

[1] T. T. Sudmann, I. T. Borsheim, K. Ovsthus, T. Ciamulski, and F. F. Jacobsen, "UWB-radar monitoring of movements in homes of elderly and disabled people. An interdisciplinary perspective (RADCARE)," in *Intelligent Data Acquisition and Advanced Computing Systems: Technology and Applications (IDAACS), 2015 IEEE 8th International Conference on*, Warzawa, 2015, vol. 2, pp. 747–750: IEEE.

[2] T. T. Sudmann, I. T. Børsheim, K. Øvsthus, T. Ciamulski, and F. F. Jacobsen, "Care support for elderly and disabled people by radar sensor technology (RADCARE)" in *The 23rd International Conference on Health Promoting Hospitals & Health Service. Person-oriented health promotion in rapidly changing world: Co-production – Continuity – New media & technologies*, Oslo, 2015, vol. 5, no. Supplement 1, p. 50: Clinical health promotion.

[3] S. L. Andersen, P. Sebastiani, D. A. Dworkis, L. Feldman, and T. T. Perls, "Health span approximates life span among many supercentenarians: compression of morbidity at the approximate limit of life span," *The Journals of Gerontology Series A: Biological Sciences and Medical Sciences,* vol. 67, no. 4, pp. 395–405, 2012.

[4] Eurostat, "Population projections 2010–2060," in "Newsrelease," 2011, Available: http://ec.europa.eu/eurostat.

[5] Eurostat, "Ageing in the European union: Where exactly?," in "Eurostat Statistics in focus," 26/2010, 2010.

[6] A. Bergland, G.-B. Jarnlo, and K. Laake, "Predictors of falls in the elderly by location," *Aging clinical and experimental research,* vol. 15, no. 1, pp. 43–50, 2003.

[7] L. Clemson, H. Kendig, L. Mackenzie, and C. Browning, "Predictors of Injurious Falls and Fear of Falling Differ An 11-Year Longitudinal Study of Incident Events in Older People," *Journal of aging and health,* vol. 27, no. 2, pp. 239–56, 2014.

[8] S. R. Lord, C. Sherrington, H. B. Menz, and J. C. Close, *Falls in older people: risk factors and strategies for prevention.* Cambridge University Press, 2007.

[9] L. Z. Rubenstein, "Falls in older people: epidemiology, risk factors and strategies for prevention," *Age and ageing,* vol. 35, no. suppl 2, pp. ii37–ii41, 2006.

[10] A. C. Scheffer, M. J. Schuurmans, N. Van Dijk, T. Van der Hooft, and S. E. De Rooij, "Fear of falling: measurement strategy, prevalence, risk factors and consequences among older persons,"*Age and ageing,* vol. 37, no. 1, pp. 19–24, 2008.

[11] J. A. Stevens, J. E. Mahoney, and H. Ehrenreich, "Circumstances and outcomes of falls among high risk community-dwelling older adults," *Inj Epidemiol,* vol. 1:5, 2014.

[12] R. Tideiksaar, *Falls in older people: prevention and management.* Baltimore: Health Professions Press, 2010, pp. xvii, 314 s.

[13] M. E. Tinetti, C. Gordon, E. Sogolow, P. Lapin, and E. H. Bradley, "Fall-risk evaluation and management: challenges in adopting geriatric care practices," *The Gerontologist,* vol. 46, no. 6, pp. 717–725, 2006.

[14] R. Rashya and M. Sindhuja, "An Enhanced & Effective Fall Detection System for Elderly Person Monitoring using Consumer Home Networks," *International journal of research in Engineering and science (IJRES),* vol. 3, no. 3, pp. 50–57, 2015.

[15] World Health Organization, *WHO global report on falls prevention in older age.* World Health Organization, 2008.

[16] S. Abbate, M. Avvenuti, P. Corsini, A. Vecchio, and J. Light, "Monitoring of human movements for fall detection and activities recognition in elderly care using wireless sensor network: a survey," in *Wireless Sensor Networks: Application-Centric Design,* G. V. Merret and Y. K. Tan, Eds. Rijeka, Croatia: Tech, 2010, pp. 147–166.

[17] *The Publich Health Report. Coping and Possibilities.*

[18] L. Fleig *et al.,* "Health behaviour change theory meets falls prevention: Feasibility of a habit-based balance and strength exercise intervention for older adults," *Psychology of Sport and Exercise,* vol. 22, pp. 114–122, 2016.

[19] L. Fleig *et al.,* ""Motivation gets you started, habit keeps you going": Feasibility of a habit-based physical activity intervention," *European Health Psychologist,* vol. 17, no. S, p. 483, 2015.

[20] N.-O. Ørjasæter and K. M. Kistorp, "Velferdsteknologi i sentrum. Innføring av velferdsteknologi i sentrumsbydelene i Oslo. En kartlegging av effekten. (Implementation of assistive technology in central Oslo. Mapping of effekt. Part 2 of 2)," Intro International & Arkitektur- og designhøgskolen i Oslo, Oslo 2016.

[21] The truth about care for next-of-kin (Når sant skal sies om pårørende omsorg).

[22] The Coordination Reform. Proper treatment – at the right place and right time (Official English version).

[23] B. Poland, P. Lehoux, D. Holmes, and G. Andrews, "How place matters: unpacking technology and power in health and social care," *Health & Social Care in the Community,* vol. 13, no. 2, pp. 170–180, 2005.

[24] World Health Organization. *Active ageing: A policy framework: World Health Organization,* 2002.

[25] B. Marent, R. Forster, and P. Nowak, "Theorizing participation in health promotion: A literature review," *Social Theory & Health,* vol. 10, no. 2, pp. 188–207, 2012.

[26] S. A. Ballegaard, "Healthcare technology in the home. Of home patients, family caregivers, and a vase of flowers," 2011.

[27] M. W. J. Schillmeier and M. Domènech, "New technologies and emerging spaces of care." Farnhamk: Ashgate, 2010.

[28] H. Thygesen, "Technology and good dementia care: a study of technology and ethics in everyday care practice," no. 159, Unipub, Oslo, 2009.

[29] M. le, C. Rocker, and A. Holzinger, "Perceived usefulness of assistive technologies and electronic services for ambient assisted living," 2011, pp. 585–592: IEEE.

[30] N.-O. Ørjasæter and K. M. Kistorp, "Velferdsteknologi i sentrum. Innføring av velferdsteknologi i sentrumsbydelene i Oslo. En kartlegging av effekten. Del 1 av 2 (Implementation of assistive technology in central Oslo. Mapping of effekt. Part 1 of 2)," Intro International & Arkitektur- og designhøgskolen i Oslo, Oslo 2015.

[31] A. Bauman, D. Merom, F. C. Bull, D. M. Buchner, and M. A. F. Singh, "Updating the Evidence for Physical Activity: Summative Reviews of the Epidemiological Evidence, Prevalence, and Interventions to Promote "Active Aging"," *The Gerontologist,* vol. 56, no. Suppl 2, pp. S268-S280, 2016.

[32] P. de Souto Barreto et al., "Recommendations on Physical Activity and Exercise for Older Adults Living in Long-Term Care Facilities: A Taskforce Report," *Journal of the American Medical Directors Association,* 2016.

[33] L. G. Johnson et al., "Light physical activity is positively associated with cognitive performance in older community dwelling adults," *Journal of Science and Medicine in Sport.*

[34] L. D. Gillespie *et al.*, "Interventions for preventing falls in older people living in the community," *Cochrane Database of Systematic Reviews Rev,* vol. 9, no. No.: CD007146, 2012.

[35] T. Masud and R. O. Morris, "Epidemiology of falls," *Age and ageing,* vol. 30, pp. 3–7, 2001.

[36] G. Baldewijns, S. Luca, B. Vanrumste, and T. Croonenborghs, "Developing a system that can automatically detect health changes using transfer times of older adults," *BMC medical research methodology,* vol. 16, no. 1, p. 1, 2016.

[37] R. Igual, C. Medrano, and I. Plaza, "Challenges, issues and trends in fall detection systems," *Biomedical engineering online,* vol. 12, no. 1, p. 66, 2013.

[38] M. Skubic *et al.*, "A framework for harmonizing sensor data to support embedded health assessment," in *Engineering in Medicine and Biology Society (EMBC), 2014 36th Annual International Conference of the IEEE,* 2014, pp. 1747–1751: IEEE.

[39] E. Boulton *et al.*, "Developing the FARSEEING Taxonomy of Technologies: Classification and Description of Technology Use (including ICT) in Falls Prevention Studies," *Journal of Biomedical Informatics,* 2016.

[40] C. Becker *et al.*, "Proposal for a multiphase fall model based on real-world fall recordings with body-fixed sensors," *Zeitschrift für Gerontologie und Geriatrie,* vol. 45, no. 8, pp. 707–715, 2012.

[41] S. N. Robinovitch *et al.*, "Video capture of the circumstances of falls in elderly people residing in long-term care: an observational study," *The Lancet,* vol. 381, no. 9860, pp. 47–54, 2013.

[42] W. Xu *et al.*, "Meta-analysis of modifiable risk factors for Alzheimer's disease," *Journal of Neurology, Neurosurgery & Psychiatry,* vol. 86, no. 12, pp. 1299–1306, 2015.

[43] T. Paillard, "Preventive effects of regular physical exercise against cognitive decline and the risk of dementia with age advancement," *Sports Medicine-Open,* vol. 2, no. 1, pp. 1–6, 2015.

[44] K. Uemura, M. Yamada, K. Nagai, B. Tanaka, S. Mori, and N. Ichihashi, "Fear of falling is associated with prolonged anticipatory postural adjustment during gait initiation under dual-task conditions in older adults," *Gait & posture,* vol. 35, no. 2, pp. 282–286, 2012.

[45] A. Shumway-Cook and M. H. Woollacott, *Motor control: translating research into clinical practice.* Philadelphia: Lippincott Williams & Wilkins, 2012, pp. XIV, 641 s. : ill.

[46] F. Pamoukdjian *et al.*, "Measurement of gait speed in older adults to identify complications associated with frailty: A systematic review," *Journal of Geriatric Oncology,* vol. 6, no. 6, pp. 484–496, 2015. doi:http://dx.doi.org/10.1016/j.jgo.2015.08.006

[47] A. Middleton, S. L. Fritz, and M. Lusardi, "Walking speed: the functional vital sign," *Journal of aging and physical activity,* vol. 23, no. 2, pp. 314–322, 2015.

[48] L. H. J. Kikkert, N. Vuillerme, J. P. van Campen, T. Hortobágyi, and C. J. Lamoth, "Walking ability to predict future cognitive decline in old adults: A scoping review," *Ageing Research Reviews,* vol. 27, pp. 1–14, 2016. doi:http://dx.doi.org/10.1016/j.arr.2016.02.001

[49] M. M. Lusardi, "Is walking speed a vital sign? Absolutely!," *Topics in Geriatric Rehabilitation,* vol. 28, no. 2, pp. 67–76, 2012.

[50] C. A. Capaldi, R. L. Dopko, and J. M. Zelenski, "The relationship between nature connectedness and happiness: a meta-analysis," (in English), *Frontiers in Psychology,* Original Research vol. 5, 2014-September-8 2014.

[51] S. Fritz and M. Lusardi, "White paper: "walking speed: the sixth vital sign"," *Journal of geriatric physical therapy,* vol. 32, no. 2, pp. 2–5, 2009.

[52] J. Klenk *et al.*, "Development of a standard fall data format for signals from body-worn sensors. The FARSEEING consensus," *Zeitschrift für Gerontologie und Geriatrie,* vol. 46, no. 8, pp. 720–726, 2013.

[53] OPAL Wearble Sensors. (29. June 2015). Available: http://apdm.com/Wearable-Sensors/Opal

[54] Qualisys Motion Capture. (29. June 2015). Available: http://www.qualisys.com/

[55] G. Abellan Van Kan *et al.*, "Gait speed at usual pace as a predictor of adverse outcomes in community-dwelling older people an International Academy on Nutrition and Aging (IANA) Task Force," *The journal of nutrition, health & aging,* vol. 13, no. 10, pp. 881–889, 2009.

[56] M. Brodie, S. Lord, M. Coppens, J. Annegarn, and K. Delbaere, "Eight weeks remote monitoring using a freely worn device reveals unstable gait patterns in older fallers," 2015.

[57] J. M. Hausdorff and A. S. Buchman, "What Links Gait Speed and MCI With Dementia? A Fresh Look at the Association Between Motor and Cognitive Function," *The Journals of Gerontology Series A: Biological Sciences and Medical Sciences,* vol. 68, no. 4, pp. 409–411, 2013.

[58] T. Ijmker and C. J. Lamoth, "Gait and cognition: the relationship between gait stability and variability with executive function in persons with and without dementia," *Gait & posture,* vol. 35, no. 1, pp. 126–130, 2012.

[59] O. Beauchet *et al.,* "Poor Gait Performance and Prediction of Dementia: Results From a Meta-Analysis," *Journal of the American Medical Directors Association,* in press.

[60] M. E. Tinetti, D. Richman, and L. Powell, "Falls efficacy as a measure of fear of falling," *Journal of gerontology,* vol. 45, no. 6, pp. P239–P243, 1990.

[61] O. A. Donoghue, H. Cronin, G. M. Savva, C. O'Regan, and R. A. Kenny, "Effects of fear of falling and activity restriction on normal and dual task walking in community dwelling older adults," *Gait & posture,* vol. 38, no. 1, pp. 120–124, 2013.

[62] E. Barry, R. Galvin, C. Keogh, F. Horgan, and T. Fahey, "Is the Timed Up and Go test a useful predictor of risk of falls in community dwelling older adults: a systematic review and meta-analysis," *BMC geriatrics,* vol. 14, no. 1, p. 14, 2014.

[63] K. W. Hayes and M. E. Johnson, "Measures of adult general performance tests: The Berg Balance Scale, Dynamic Gait Index (DGI), Gait Velocity, Physical Performance Test (PPT), Timed Chair Stand Test, Timed Up and Go, and Tinetti Performance, ÄêOriented Mobility Assessment (POMA)," *Arthritis Care & Research,* vol. 49, no. S5, pp. S28–S42, 2003.

[64] T. M. Steffen, T. A. Hacker, and L. Mollinger, "Age-and gender-related test performance in community-dwelling elderly people: Six-Minute Walk Test, Berg Balance Scale, Timed Up & Go Test, and gait speeds," *Physical therapy,* vol. 82, no. 2, pp. 128–137, 2002.

[65] A. Yingyongyudha, V. Saengsirisuwan, W. Panichaporn, and R. Boonsinsukh, "The Mini-Balance Evaluation Systems Test (Mini-BESTest) Demonstrates Higher Accuracy in Identifying Older Adult Participants With History of Falls Than Do the BESTest, Berg Balance Scale, or Timed Up and Go Test," *Journal of Geriatric Physical Therapy,* vol. 39, no. 2, pp. 64–70, 2016.

[66] D. Webster and O. Celik, "Systematic review of Kinect applications in elderly care and stroke rehabilitation," *Journal of Neuroengineering and Rehabilitation,* vol. 11, no. 108, pp. 1–24, 2014.

[67] G. Baldewijns *et al.,* "Fall prevention and detection," in *Active and assisted living. Technologies and applications,* F. Florez-Revuelta and A. A. Chaaraouni, Eds. London: Institution of Engineering and Technology, 2016, pp. 203–223.

[68] M. M. Lusardi, "Using walking speed in clinical practice: interpreting age-, gender-, and function-specific norms," *Topics in Geriatric Rehabilitation*, vol. 28, no. 2, pp. 77–90, 2012.

[69] D. M. Peters, S. L. Fritz, and D. E. Krotish, "Assessing the reliability and validity of a shorter walk test compared with the 10-meter walk test for measurements of gait speed in healthy, older adults," *Journal of Geriatric Physical Therapy*, vol. 36, no. 1, pp. 24–30, 2013.

[70] S. Studenski, "Bradypedia: Is gait speed ready for clinical use?," *The journal of nutrition, health & aging*, vol. 13, no. 10, pp. 878–880, 2009.

[71] V. J. Verlinden, J. N. van der Geest, A. Hofman, and M. A. Ikram, "Cognition and gait show a distinct pattern of association in the general population," *Alzheimer's & Dementia*, vol. 10, no. 3, pp. 328–335, 2014.

[72] D. Levine, J. Richards, and M. W. Whittle, *Whittle's gait analysis*. Elsevier Health Sciences, 2012.

[73] S. Studenski *et al.*, "Gait speed and survival in older adults," *Jama*, vol. 305, no. 1, pp. 50–58, 2011.

[74] R. W. Bohannon and A. W. Andrews, "Normal walking speed: a descriptive meta-analysis," *Physiotherapy*, vol. 97, no. 3, pp. 182–189, 2011.

[75] J. Wagner and R. Z. Morawski, "Estimation of trajectories," Institute of Radioelectronics and Multimedia Technology, Warsaw University of Technology, internal report 2015.

[76] Mathworks. (2016-02-17). *Matlab 2015R Documentation: atan2 – Four-quadrant inverse tangent*. Available: http://www.mathworks.com/help/matlab/ref/atan2.html

[77] Mathworks. (2016-02-07). *Matlab 2015R Documentation: unwrap – Correct phase angles to produce smoother phase plots*. Available: http://www.mathworks.com/help/matlab/ref/unwrap.html

[78] E. Håland and L. Melby, "Negotiating technology-mediated interaction in health care," *Social Theory & Health*, vol. 13, no. 1, pp. 78–98, 2015.

[79] L. Magnusson and E. J. Hanson, "Ethical issues arising from a research, technology and development project to support frail older people and their family carers at home," *Health & Social Care in the Community*, vol. 11, no. 5, pp. 431–439, 2003.

[80] T. Greenhalgh and D. Swinglehurst, "Studying technology use as social practice: the untapped potential of ethnography," *BMC medicine*, vol. 9, no. 1, p. 45, 2011.

[81] J. Van Hoof, H. Kort, P. Rutten, and M. Duijnstee, "Ageing-in-place with the use of ambient intelligence technology: Perspectives of older users," *International Journal of Medical Informatics*, vol. 80, no. 5, pp. 310–331, 2011.

[82] M. Ziefle, C. Rocker, and A. Holzinger, "Medical Technology in Smart Homes: Exploring the User's Perspective on Privacy, Intimacy and Trust," 2011, pp. 410–415: IEEE.

[83] A. Kiran, "The primacy of action. Technological co-constitution of practical space," Doctoral Thesis Norwegian University of Science and Technology 2009: 105, 2009.

[84] C. Shilling, "The body in culture, technology and society," ed. London: SAGE, 2005.

[85] World Health Organisation, "World report on ageing and health," ed. Luxenburg: World Health Organisation, 2015.

[86] L. Fleig *et al.*, "Sedentary Behavior and Physical Activity Patterns in Older Adults After Hip Fracture: A Call to Action," *Journal of aging and physical activity,* vol. 24, no. 1, pp. 79–84, 2016.

[87] J. Hamm, A. G. Money, A. Atwal, and I. Paraskevopoulos, "Fall prevention intervention technologies: A conceptual framework and survey of the state of the art," *Journal of Biomedical Informatics,* 2016.

[88] Y. S. Delahoz and M. A. Labrador, "Survey on fall detection and fall prevention using wearable and external sensors," *Sensors,* vol. 14, no. 10, pp. 19806–19842, 2014.

2

A System for Elderly Persons Behaviour Wireless Monitoring

Jerzy Kołakowski[1], Magdalena Berezowska[1], Ryszard Michnowski[1], Karol Radecki[1] and Lukasz Malicki[2]

[1]Institute of Radioelectronics, Nowowiejska 15/19, 00-665 Warsaw, Poland
[2]Knowledge Society Association, Grazyny 13/15 lok. 221,
02-548 Warsaw, Poland

Abstract

This chapter contains a description of the measurement system intended for gathering data typically used for the evaluation of elderly person's mobility and behaviour. The measurements are carried out with micro-electro-mechanical systems sensors employed in the mobile device worn by the monitored person. Results of measurements are sent over wireless link to recording nodes for analysis. This chapter describes system design and presents examples of recorded data. Proposals of system usage for mobility and behaviour investigation are also presented and discussed.

Keywords: Elderly persons, Health care, Behaviour investigation.

2.1 Introduction

In the European population, the ageing people tend to live longer and more independently. Many efforts are being put into development of systems monitoring every day in elderly person's activities and behaviour, in order to provide good health and a good quality of life at home.

Recently, different hardware and software platforms have been developed for this purpose. The developed solutions can be grouped into two categories:

- wearable devices, and
- non-wearable solutions.

Wearable devices usually utilize different sorts of sensors. They can easily detect body movements and measure vital signs. They should be worn by the monitored person what is perceived as a drawback.

Non-wearable solutions are less intrusive. They are usually installed at elderly person's home. A wide selection of devices can be used for monitoring purposes: audio devices or infrared sensors for detection of person presence, pressure sensors placed under floor mats or sensors measuring floor vibration for recording person steps and sensors measuring electric energy consumption are only few examples [1]. Data fusion of results gathered by such multisensor system allows for non-intrusive elderly person behaviour evaluation. The drawbacks of such solutions are lower reliability and accuracy than in case of wearable devices.

The vast majority of solutions presented in the literature are based on Micro-Electro-Mechanical Systems (MEMS) sensors, usually incorporated into devices worn by elderly persons [2–5]. They are used to collect data from individual body segments over extended periods of time.

The most popular sensors, accelerometers, are typically used for characterization of postures and activities. Common electronic devices (e.g., cellular phones) equipped with accelerometers and worn on the wrist or ankle, are currently widely used to record activity over long time period.

Accelerometers are commonly used for monitoring daily activity patterns, including gait, sit-to-stand transfers, postural patterns, and falls [6–9]. Falls are the most common type of home accidents among elderly people and can affect their health and independence [10]. In a number of studies wearable devices based on accelerometers have been used for fall risk evaluation and fall detection [11–14].

In order to increase system efficiency and reliability data, fusion of results from many different sensors can be used. The typical choice of sensors comprises accelerometers, gyroscopes, and magnetic or atmospheric pressure sensors. Examples of such designs can be found in the references Jia et al. [15] and Barth and Brown [16].

Triaxial gyroscopes and magnetometers provide the most accurate measurements of angular orientation during movement [17, 18]. Addition of altitude information from the barometric pressure sensor significantly increases the activity classification accuracy.

Elderly persons mobility and behaviour investigation is one of NITICS (InfrasTructure for Innovative home Care Solutions) Ambient Assisted Living (AAL) project objectives [19]. The system described in this chapter is a part of the NITICS platform gathering information on the elderly person's health status. The platform architecture is shown in Figure 2.1.

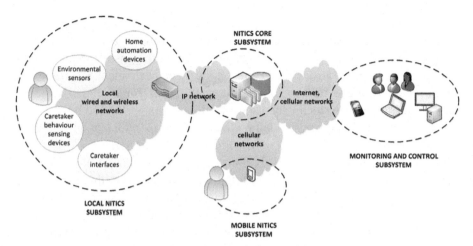

Figure 2.1 NITICS platform architecture.

The NITICS system is composed of four subsystems. NITICS Core Subsystem (NCS) is a component that carries out system control and database functions. It stores user profiles and measurement results were delivered by subsystems. It is responsible for the implementation of NITICS services.

Local NITICS Subsystem (LNS) comprises devices installed in caretaker's home and communication with the NCS via single local fixed (LAN) or wireless (WLAN, cellular) link. The LNS controls devices and sensors, acquires measurement results and transfers them to the NCS. Similar functions are carried out by a Mobile NITICS Subsystem (MNS), but the components of the MNS are worn by the caretaker. This allows NITICS system to operate outdoor in places not covered by the LNS.

Monitoring and Control Subsystem (MCS) comprises equipment used by caregivers and system management staff. Its functions include: service control, service data management, and presentation of data to caregivers.

The system described in this chapter is a component of Local NITICS Subsystem. The system collects data that can be used for:

- immediate alert generation in case of a sudden change of behaviour,
- analysis of a long-term changes in person's mobility which can help predict and identify abnormal behaviours (e.g., continuously decreasing physical activity can be a symptom of approaching mobility problems, fast reaction to the problem can slow down or stop the process).

Similar functionalities can be found in several solutions described in the literature. Majority of them relies on sensors and communication capabilities of the smartphone. Example of such approach is described in He and Li [20]. Some solutions use external sensors and smartphone for processing gathered results. Vigilante [21] system gathers data with sensors located in a bracelet. Results are transferred to the phone over permanent Bluetooth link. Similar solution was proposed in Cosar system [22]. Smartphone-based systems are usually able to locate the user. Localization is performed by GPS, cellular triangulation system (provided by the network operator), or an Radio Frequency Identification (RFID) systems (indoor applications).

The solution proposed in the NITICS consists in development of devices that do not rely on the smartphone functionality [23]. Thanks to that the tags are smaller and can be embedded in person clothes. The system operates in 868 Hz band, where propagation conditions are better. Therefore, the system coverage and reliability is higher. Besides sensors data characterizing monitored person movements, the system estimates person location which can support the behaviour evaluation.

2.2 System for Mobility Investigation

2.2.1 System Components

A system dedicated to mobility data recording comprises the following devices and accessories:

- three recording nodes intended for measurement results storage (nodes are equipped with microSD cards, 868 MHz transceivers, and WiFi modules),
- one mobile node equipped with sensors for mobility parameter measurements,
- accessories (e.g., mobile and recording nodes chargers, accessories for mobile sensor attachment).

A typical measurement scenario is shown in Figure 2.2. The recording nodes are distributed in an elderly person's home. The mobile node attached to the belt is worn by the elderly person. The node measures parameters that can be used for movement characterization (acceleration, angular rate, magnetic field, and atmospheric pressure). Results are buffered and periodically broadcasted. Recording nodes receive the messages, store the results, and optionally transmit them to the PC computer for further processing.

Additionally, recording nodes measure the levels of RF signals reaching their receivers. The results are stored together with obtained messages. The information on signal level can be used to estimate the distance between the

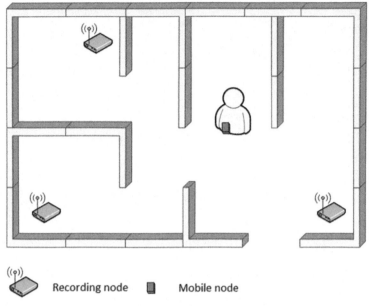

Figure 2.2 System usage scenario.

mobile and recording nodes which can be the basis for estimation of the mobile node location. Information on location can support analysis of recorded results and can be also used for the elderly person behaviour evaluation.

The mobile node block diagram is shown in Figure 2.3. The node operation is controlled by the MSP430FR5739 microcontroller, where the program is stored in its FRAM memory. This technology significantly reduces current consumption and increases number of allowed write–read operations. The chip uses SPI interfaces to communicate with two MEMS sensors (BMX 055 and BMP183 both from Bosch Sensortec) and the UART interface for wireless communication module (XBee 868LP from Digi Inc.) control.

The module operates in 868-MHz frequency band and provides data transmission with 80 kbit/s rate. The module can be used for point to point transmission and also can broadcast messages to all nodes in the mobile node vicinity.

The recording node is responsible for the following tasks:

- reception of results from the mobile node,
- measurement of RF signal level,
- storage of received and measured data on the microSD card, and
- sending the results over WiFi link to the external PC.

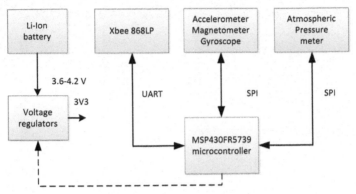

Figure 2.3   Mobile node block diagram.

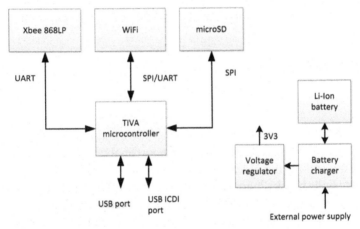

Figure 2.4   Recording node block diagram.

The recording node block diagram is presented in Figure 2.4. Node operation is controlled by the TIVA family (TM4C123GH6PMI) ARM microcontroller. It is responsible for controlling the XBee wireless module and the microSD card. The node is equipped with two Li-Ion batteries. On the node PCB a battery charger chip is assembled, so it can also operate with external power supply. Pictures of recording and mobile nodes are presented in Figure 2.5.

2.2.2 System Operation

After switching the recording node on, the initialization of the microcontroller is carried out. Next, a wireless communications module (the same module as

Figure 2.5 View of recording and mobile nodes.

in the mobile node was used) is initialized and the result file is created on the micro-SD card. The file is opened in append mode. After writing 100 messages, the file is closed and the new one is opened.

After correct initialization of all recording nodes, the mobile node should be switched on. The node starts operation from initialization of microcontroller internal components (clock oscillators, ports, interfaces, etc.), and then the XBee module and both MEMS chips are configured.

The mobile node operation is controlled by timers. They periodically trigger interrupts resulting in awaking of the microcontroller and on-board sensors. After performing their tasks the tag components are turned into sleep modes. MEMS sensors results are temporarily stored in the microcontroller memory and are broadcasted over the 868 MHz radio link to recording nodes twice per second. Sampling frequencies, measurement ranges and resolutions, characterizing measurements carried out with MEMS sensors, are presented in Table 2.1.

The results are grouped into frames and sent to recording nodes. During reception of the frame each recording node measures received RF signal level. The results are transferred in binary frames. Each frame is equipped with a unique identifier ranging from 0 to 65536. After each transmission, the identifier number is incremented. Switching the mobile node off and on again, resets this value. Frames received over radio link are appended with corresponding RF signal level value and finally written to the file.

Table 2.1   Data acquired with MEMS sensors

MEMS Sensor	Sampling Frequency (Hz)	Measurement Range	Resolution
Accelerometer	20	± 8 g	0.98 mg
Gyroscope	20	± 2000 deg./s	0.004 deg.$^\circ$/s
Magnetometer	10	± 1200 μT (x, y) and ± 2500 μT (z)	0.3 μT
Atmospheric pressure	2	$300 \ldots 1100$ hPa	2 Pa (0.17 m)

2.3 Test Campaign

Test campaign consisted of five all-day measurement sessions, carried out in elderly person's homes. All men participating in tests were 65 years old and over. Each measurement was preceded by a calibration procedure consisting in performing measurements in all rooms. Gathered results were used for tuning localization procedures. After the calibration, the elderly person received the mobile node and attached it to his belt. The evaluator asked him not to change his habits during the test-day.

Exemplary results gathered during test campaign are presented in Figures 2.6–2.8. The plots correspond to walking down the stairs activity. Increase in atmospheric pressure is a result of altitude change close to 3 m. Abrupt changes of acceleration are caused by steps. Each step resulted in peaks in the acceleration plot (Figure 2.7) Additional information, giving more details to of movement description can be extracted from the results of measurements taken with the magnetometer. Two changes in the movement direction corresponding to stairs, with a landing in the middle can be noticed. During walk also a small changes of magnetic field can be observed (Figure 2.8). They are caused by body swing during movement between consecutive steps.

2.4 Results Analysis

Collected measurements results can be a basis for evaluation of elderly person mobility and behaviour. The approach to result processing developed within the NITICS project assumes evaluation of the following parameters:

- elderly person activity analysis (gait period distribution, total time of walking, and distribution of walking time over a day), and
- room occupancy.

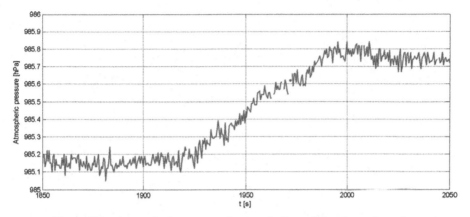

Figure 2.6 Atmospheric pressure changes during walking down the stairs.

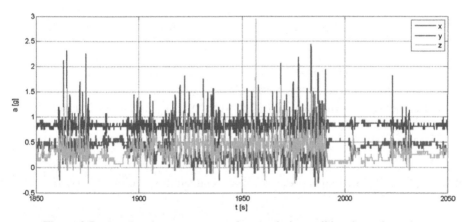

Figure 2.7 Acceleration components changes during walking down the stairs.

2.4.1 Activity Analysis

Identification of the gait is crucial for the evaluation of two first parameters. The algorithm developed within the project is based on wavelet processing of accelerometer sensor results. Wavelet transform is a perfect tool to analyse non-stationary signals. Parts of recorded signals corresponding to walking periods are transferred to the time-scale plane with continuous wavelet transform. The moments of particular steps are determined from transform coefficients modules.

Wavelet transform of $s(t)$ signal can be expressed as [24]:

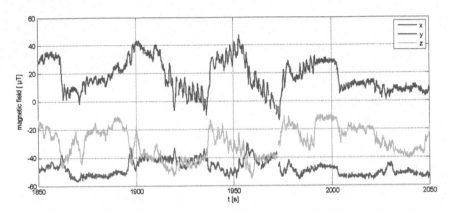

Figure 2.8   Magnetic field components changes during walking down the stairs.

$$W_s(a, b) = \int\limits_{-\infty}^{\infty} s(t)\psi_{a,b}{}^*(t)\mathrm{d}t, \tag{2.1}$$

where $s(t)$ – transformed signal,

$\psi_{a,b}{}^*(t)$ – the complex conjugate wavelet function.

The wavelet is a mother wavelet function scaled by a and shifted in time by b

$$\psi_{a,b}(t) = \frac{1}{\sqrt{a}}\psi\left(\frac{t - b}{a}\right). \tag{2.2}$$

In proposed algorithm fourth order complex Gaussian wavelet was used:

$$\psi(t) = C\frac{\mathrm{d}^4}{\mathrm{d}t^4}\left(e^{-jt}e^{-t^2/2}\right), \tag{2.3}$$

where C is a normalization constant such that

$$\|\psi(t)\| = 1. \tag{2.4}$$

The mother wavelet's real and imaginary parts are shown in Figure 2.9.

Wavelet transform module is a time-scale signal representation. Higher values of wavelet transform coefficients have greater amplitudes in the vicinity of time values corresponding to greater changes of acceleration. The wavelet transform simplifies gait detection because of its de-noising properties. Coefficients corresponding to the signal are grouped, but noise coefficients are spread over the time-scale plane.

Figure 2.10 shows an example of recorded signal, its wavelet transform module and wavelet transform coefficient module for scale (a) equal to 3.

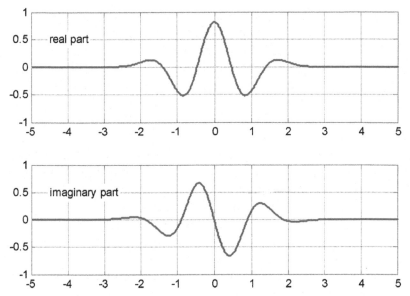

Figure 2.9 Real and imaginary parts of the fourth-order complex Gaussian wavelet.

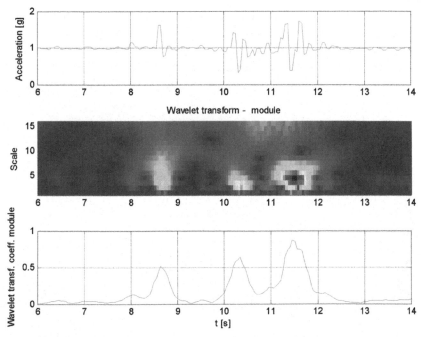

Figure 2.10 Recorded signal – its wavelet transform module and wavelet transform coefficient module for scale equal to 3.

Gait period determination relies on location of wavelet coefficient module peaks. The peaks correspond to changes of acceleration, caused by steps. Peak position determination procedure has two parameters, which should be tuned to particular person gait: peak threshold and peak separation. The first one eliminates low peaks resulting from slight body movements, the second eliminates interfering peaks which are too close to the primary peak (such situation is observed when acceleration change caused by step is stronger and additional ringing on the acceleration graph is observed).

Collected accelerometer data along with previously mentioned timestamps allow to roughly separate sensor data samples corresponding to the gait from data related to any other aperiodic activity. The proposed algorithm is based on the detection of acceleration peaks related to heel-strike events.

In case of a pair of peaks situated too close to each other (in distance of less than step window), algorithm chooses dominant peak. Sample corresponding to detected peak is interpreted as step event moment.

Identification of particular steps allows to calculate gait period which can be a basis for other parameters evaluation. Exemplary gait period histogram is presented in Figure 2.11(a). It was obtained from data gathered during one day measurement session.

Analysis of data acquired during test campaign showed that histograms drawn for people with mobility problems are different from ones obtained for other monitored persons.

Another approach to gait evaluation can be based on cumulative distribution function (CDF), calculated for all day elderly person walking activities.

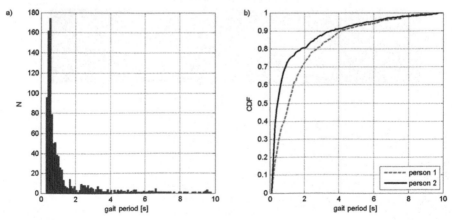

Figure 2.11   Gait period histogram (a), comparison of CDFs obtained for two monitored persons (b).

Comparison of CDF functions calculated for two monitored persons is shown in Figure 2.11(b). Mobility problems (one of monitored persons was hobbling) resulted in moving the CDF curve to the right.

The idea of testing depends on the development of the CDF function mask. Crossing the mask may be a symptom of problems with mobility.

Changes in elderly person activity can be a symptom of illness or deterioration in wellbeing. Two parameters were used as activity measures:

- total time of walking defined as a sum of walking periods. Walking period is a sum of time intervals between consecutive steps. It was assumed that time intervals greater than 10 s are considered as a break in activity; and
- walking period distribution showing the time that the elderly person spent on walking during each hour a day.

Both parameters allow for detection of abnormal monitored person behaviour. Such data can be used in long-term analysis of the elderly person behaviour or to trigger an alert, if the activity is much lower than expected.

2.4.2 Room Occupancy Determination

Information on RF signal levels was used for the estimation of monitored person location. This information allows for the determination of room occupancy defined as a percentage of time spent by the elderly person in particular rooms. Usage of three or more recording nodes allows to estimate this parameter.

The exemplary graphs showing changes of RF signal levels received by three recording nodes installed in different rooms are presented in Figure 2.12. Abrupt changes in signal level are mainly caused by the multipath propagation channel. The signal transmitter was attached at waist level. Therefore, in most cases it operated in NLOS (Non-Line-of-Sight) conditions. The direct paths between transmitter antenna and recording nodes were obscured by the person body, furniture, and room walls.

The RF signal level allows for estimation of distances between the transmitter and receiving nodes. In implemented algorithms simple channel model was taken. The distance estimate d is equal to:

$$d = 10^{\frac{A-RSSI}{10n}},\qquad(2.5)$$

where: A signal level recorded at 1m distance,
RSSI Received signal strength indicator, and
n propagation constant.

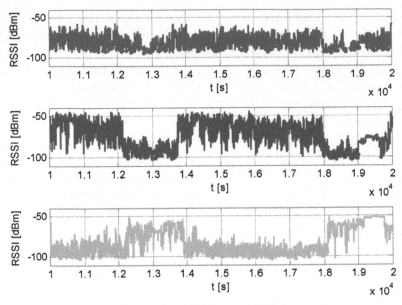

Figure 2.12   RF signal levels recorded by three nodes.

Before start of the measurement a short calibration procedure was performed. It consisted in measurement of signal levels at several test points in the elderly person flat. Recorded results were used for determination of A level. The averaged value for results presented below was equal to –55 dBm. All gathered results were used for determination of propagation constant n. Calculated distance MSE criterion was used for n selection (n equal to 3.5 was taken in considered case).

A set of calculated distances was used for determination of monitored person location. The positioning algorithm based on the Kalman filter was implemented. Although during the tests a system including three recording nodes was tested, the proposed algorithm performs well also in over-determined systems, where larger number of recording nodes is used.

The current position is described by a state vector including x, y coordinates and components of the mobile node velocity.

$$x_k = [x \ y \ v_x \ v_y]^T. \tag{2.6}$$

The Kalman filter operation consists in successive prediction and correction of the state vector. In the correction phase results of distance evaluation, based on RF signal level measurements and described propagation channel model is performed.

Distance at k-th moment between mobile node and m-th recording node is equal to

$$d_{mk} = \sqrt{(x_k - x_m)^2 + (y_k - y_m)^2 + (z - z_m)^2}. \qquad (2.7)$$

Since the mobile node was mounted at waist level z was assumed to be constant.

The measurement model is non-linear. Therefore, the Extended Kalman Filter version [25] was chosen. The measurement function h_k and Jacobian H_k matrices have the following form:

$$h_k = \begin{bmatrix} d_{1k} \\ d_{2k} \\ \dots \\ d_{Nk} \end{bmatrix} \quad H_k = \begin{bmatrix} \frac{\partial d_{1k}}{\partial x_k} & \frac{\partial d_{1k}}{\partial y_k} & 0 & 0 \\ \frac{\partial d_{2k}}{\partial x_k} & \frac{\partial d_{2k}}{\partial y_k} & 0 & 0 \\ \dots & \dots & \dots & \dots \\ \frac{\partial d_{Nk}}{\partial x_k} & \frac{\partial d_{Nk}}{\partial y_k} & 0 & 0 \end{bmatrix}.$$

Before each test the monitored person was asked to make a short walk around the flat and visit the places where recording nodes are located. The positioning results were used for simple check of the propagation channel n parameter choice correctness. The results gathered during such a walk are shown in Figure 2.13. Location of recording nodes is marked with stars. Blue dots correspond to calculated positions.

The results show that the algorithm gives results lying mostly in the flat area. However, due to low-positioning accuracy tracking of the person moving between rooms is not possible. Another results recorded during one of tests are presented in Figures 2.14 and 2.15. There is no problem with assigning the person to particular rooms.

Location information obtained from measured signal levels is not accurate because of strong impact of multipath propagation on the results. Positioning accuracy can be improved by increasing the number of recording nodes or by implementation of other radio technologies more suited for localization, e.g., UWB. The advantage of chosen solution is a relatively small carrier frequency (868 MHz), providing very good indoor communication range.

Results gathered during the test were used for determination of room occupancy. Corresponding graphs are presented in Figure 2.16. Numbers of visits in particular rooms and time of visits can be easily read.

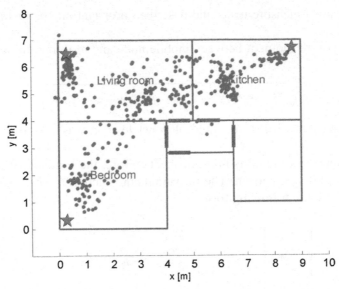

Figure 2.13   Positioning results – monitored person visited rooms with recording nodes installed.

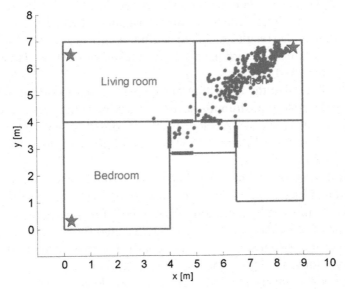

Figure 2.14   Positioning results – monitored person works in the kitchen.

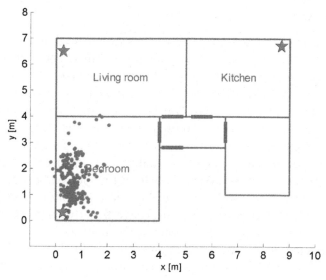

Figure 2.15 Positioning results – monitored person rests in the bedroom.

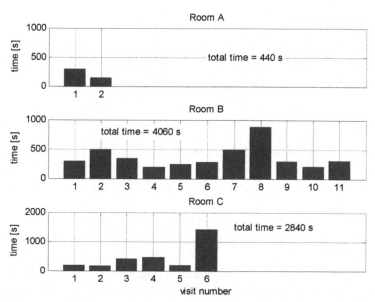

Figure 2.16 RF signal levels recorded by three nodes (a) Room occupancy graphs (A, bedroom; B, living room; and C, kitchen) (b).

2.5 Conclusion

This chapter contains a description of the wireless system used for collecting of measurement results describing elderly person's mobility. The mobile node, worn by the monitored person is equipped with MEMS sensors measuring acceleration, magnetic field, atmospheric pressure, and rotation rate.

The unique feature of the system consists in the transmission of results over radio interface. Signals are transmitted in the 868-MHz frequency band what provides system coverage sufficient for proposed application. Use of a few recording nodes increases system reliability. Results of RF signal level measurements performed by the nodes allow for estimation of the room where person's activity takes place. Although the RSSI-based positioning technique is not accurate, obtained location information significantly supports gathered data analysis and can be used for room occupancy analysis.

Results collected during the test campaign with participation of elderly persons are a valuable basis for further research on algorithms for mobility and behaviour characterization.

Acknowledgements

The research leading to these results has received funding from the National Centre for Research and Development under Grant Agreement AAL5/3/2013 (NITICS project).

References

[1] Debes, C., Merentitis, A., Sukhanov, S., Niessen, M., Frangiadakis, N., and Bauer, A. (2016). Monitoring Activities of Daily Living in Smart Homes: Understanding human behavior. *IEEE Signal Proc. Mag.* 33, 81–94.

[2] Lara, O. D., and Labrador, M. (2013). A survey on human activity recognition using wearable sensors. *IEEE Commun. Surveys Tutorials* 15, 1192–1209.

[3] Khan, A. M. et al. (2013). "Wearable recognition system for physical activities," in *9th International Conference on Intelligent Environments* (Athens, Greece), 245–249.

[4] Huynh, D. T. (2008). *Human Activity Recognition with Wearable Sensors.* PhD thesis, Technische Universität Darmstadt, Darmstadt.

[5] Parkka, J. et al. (2006). Activity classification using realistic data from wearable sensors. *IEEE Trans. Inform. Technol. Biomed.* 10, 119–128.

[6] Mathie, M. J. et al. (2004). Accelerometry: providing an integrated, practical method for long-term, ambulatory monitoring of human movement. *Physiol. Measure.* 25, R1–R20.

[7] Jia, N. (2009). Detecting human falls with a 3-axis digital accelerometer. *Analog Dialogue* 43-07, 1–7.

[8] Wang, Z. et al. (2014). A system of human vital signs monitoring and activity recognition based on body sensor network. *Sensor Rev.* 34, 42–50.

[9] Muro-de-la-Herran, A., Garcia-Zapirain, B., and Mendez-Zorrilla, A. (2014). Gait analysis methods: an overview of wearable and non-wearable systems, highlighting clinical applications. *Sensors*, 14, 3362–3394.

[10] Mendulkar, A., et al. (2014). A Survey on efficient human fall detection system. *Int. J. Sci. Technol. Res.* 3, 96–98.

[11] Bianchi, F., et al. (2010). Barometric Pressure and triaxial accelerometry-based falls event detection. *IEEE Trans. Neural Syst. Rehabil. Eng.* 18, 619–627.

[12] Kulkarni, S., and Basu, M. (2013). A review on wearable tri-axial accelerometer based fall detectors. *J. Biomed. Eng. Technol.* 1, 36–39.

[13] Aguiar, B., et al. (2014). "Accelerometer-Based Fall Detection for Smart-phones," in *IEEE International Symposium on Medical Measurement and Application* (Rome: IEEE), 1–6.

[14] Wang, J. et al. (2014). An enhanced fall detection system for elderly person monitoring using consumer home networks. *IEEE Trans. Cons. Electron.* 60, 23–29.

[15] Jia, F. et al. (2013). A home monitoring system for elderly people based on mems sensors and wireless networks. *Sensors*, 1–4.

[16] Barth, A. T., and Brown, C. L. (2009). "TEMPO 3.1: A Body Area Sensor Network Platform for Continuous Movement Assessment," in *IEEE Sixth International Workshop on Wearable and Implantable Body Sensor Networks* (Rome: IEEE), 71–76.

[17] Roetenberg, D., et al. (2005). Compensation of magnetic disturbances improves inertial and magnetic sensing of human body segment orientation. *IEEE Trans. Neural Syst. Rehabil. Eng.* 13, 395–405.

[18] Luinge, H. J., and Veltink, P. H. (2005). Measuring orientation of human body segments using miniature gyroscopes and accelerometers. *Med. Biol. Eng. Comput.* 43, 273–282.

[19] NITICS web site. Available at: http://nitics.eclexys.com/

[20] He, Y., and Li, Y. (2013). *Physical Activity Recognition Utilizing the Built-in Kinematic Sensors of a Smartphone*. Cairo: Hindawi Publishing Corporation.

[21] Lara, O. D., and Labrador, M. A. (2012). "A mobile platform for real time human activity recognition," in *Proceedings of IEEE Conference on Consumer Communications and Networks* (Rome: IEEE). doi: 10.1109/CCNC.2012.6181018

[22] Riboni, D., and Bettini, C. (2011). Cosar: hybrid reasoning for context-aware activity recognition. *Pers. Ubiquitous Comput.* 15, 271–289.

[23] Kolakowski, J., Berezowska, M., Michnowski, R., Radecki, K., Malicki, L. (2015). "Wireless system for elderly persons mobility and behaviour investigation," in *IEEE 8th International Conference on Intelligent Data Acquisition and Advanced Computing Systems: Technology and Applications (IDAACS)*, Vol. 2 (Warsaw: IEEE), 833–837.

[24] Mallat, S. (2008). *A Wavelet Tour of Signal Processing: The Sparse Way*, 3rd Edn. Cambridge, MA: Academic Press.

[25] Grewal, M. S., and Andrews, A. P. (2008). *Kalman Filtering: Theory and Practice Using MATLAB*, 3rd Edn. Hoboken, NJ: Wiley.

3

Polychromatic LED Device for Measuring the Critical Flicker Fusion Frequency

Alexey Lagunov, Ludmila Morozova, Dmitry Fedin,
Nadejda Podorojnyak, Vladimir Terehin and Aleksander Volkov

Northern (Arctic) Federal University named after M. V. Lomonosov,
163002 Severnaya Dvina Emb. 17, Arkhangelsk, Russia

Abstract

The chapter deals with a device for determining the critical flicker fusion frequency (CFFF). CFFF method is used in ophthalmology and physiology as an indicator of functional liability of the retina and visual pathway. Based on the analysis of available solutions, authors describe the creation of a new instrument. The main advantages of the device are the following: small size of glasses and instruments, computer control of the process, and availability to upload new firmware into the device by means of software. The device was tested in Nothern (Arctic) Federal University named after M.V. Lomonosov (NArFU).

Keywords: Critical flicker fusion frequency (CFFF), LED glasses, Terminal.

3.1 Introduction

During ocular organ research, flickering light source was firstly used by Bellyarminov [1]. Later, experiments showed that visual fatigue decreases critical flicker fusion frequency (CFFF). Under CFFF, we understand the minimum frequency of light interruptions per second which the intermittent light stops seeming to be flickering at, but it makes an impression of smooth light without changing brightness [2]. Dantsig, Shubova, and Mkrtycheva proved the sensitivity of the CFFF method for visual fatigue degree assessment [3].

Ohremenko's work [4] gives results of visual fatigue assessment research of people doing precision work (diamond cutter). CFFF is noted to decrease by 31.9 ± 5.9% at the end of the shift.

Vinogradov [5] presented the fact that the most typical feature of human organism fatigue is CFFF decreasing as the whole. Later on, Gavriyski, Dushkov, and Shopov found out that when size and intensity of the CFFF changes under physical training, CFFF monitoring can assess the functional state of organism, its fatigue degree [3, 6]. According to Peshkov [7], CFFF rate along with that of dynamometer, tapping test, and electrodermal resistance are the most informative criteria of the physical condition of the human body.

In the Arctic sea medicine [8], CFFF rate is used more frequently than other methods of research because of such advantages as ease of technique, portable equipment, minor time costs, and high information capability in determining fatigue of a body. The conducted researches suggested that the rate of CFFF was convenient, and revealing as a criterion of fatigue, it characterises objectively the dynamics of the health and development of body fatigue in the process of watching and during the trip [8]. In occupational medicine, the CFFF method is used along with other psycho-physiological, clinical, and physiological methods [9].

Available data about CFFF rate dependence on the human body fatigue degree are understandable from the point of view of the theory of fatigue. It was found that in the development of fatigue, caused by physical or mental work, the main role belongs to the central nervous system. According to Rosenblatt's [10] research, human body fatigue is the whole process with the Central cortical leading element representing the biological nature of cortical defensive reaction, and in the physiological mechanisms, it represents primarily working capacity decrease of cortical cells that is due to their protective inhibition. Given that CFFF rate is determined by higher departments of the visual analyser (since the central visual neurons and visual cortex are the most inert functioning of the visual system [2, 3]) when there is body fatigue due to lowering of the cortical cells efficiency, CFFF rate decreases that allows you to control the functional condition of a body and its fatigue degree according to CFFF change.

If flicker occurs with low frequency, the person sees separate flashes of light. With increasing the frequency of flashes, a person feels like he sees flickering. Thus, increasing the frequency of flashes one can get a picture where individual flashes of light are perceived by an eye as a continuous (without interruptions) light. Determination of the CFFF is performed to assess the state of the visual pathways and functional lability of the retina. Field-of-use of the study: ophthalmology, neurology, and psychiatry.

The patient is asked to look at the flickering light source and to tell a doctor when light begins to be constant for him/her and not interrupted and flickering. The frequency at which the flash point is shown with in a marked portion by the patient moment is rated as CFFF. Modern equipment allows us to investigate CFFF not only for white colour, but also for blue, red, and green ones. For exact results, the study is repeated three times for each eye.

Critical flicker fusion frequency norm for a healthy person is 40–46 Hz. CFFF reduction happens according to the age of human organism, norm of older human rarely exceeds 38–40 Hz. The value of CFFF for two healthy eyes is usually the same; the maximum difference is 5–8 Hz. Within one field of vision, CFFF parameters may vary: on the periphery in the nasal and temporal fields it is 10–15 Hz above than in the area of the macule.

Critical flicker fusion frequency depends on many factors, such as brightness and the test field size, its place on the retina projection, the spectral composition of light, length and depth of modulation stimuli, and their number after repeated presentation. Most researchers use photoimpact with duration of light stimulus that are half of their presentation cycle in length. However, CFFF measurement error depends on the duration of the light stimulus. It should be noted that CFFF depends not only on the method of measurement, but also on the physiological state of a person. CFFF does not depend on visual acuity and refraction, and it characterises the functional state of the visual analyser on the whole.

Critical flicker fusion frequency method is a subjective psychophysical one. To get objective flickers perception measures it needs to have equipment that allows generating measuring signals in wide frequency range with possibility to create different colour gamut.

3.2 Colour Vision Theories

Colour vision serves as an information source and as an element of general adaptation ability to percept a structure of surfaces and objects around us. The perception of light is determined with the wavelength of light stimulating the visual system. The range of wavelengths capable of causing human colour sensation is from 380 nm to 760 nm. Colour perception is a subjective result of influencing on the visual system of the reflected beam belonging to the visible spectrum and having a specific wavelength. That is, colour depends on how the visual system interprets the light rays of different wave length reflected from the object surface.

Perception of colour begins when a photon having fallen on the retina triggers a complex chain of biochemical transformations of photosensitive pigments. As a result, under the influence of photochemical processes, membrane potential of photoreceptors being supported as permanent the entire period of the light action changes, at the same time a photoreceptor hyperpolarises. The decrease in light causes a decrease of membrane potential and depolarisation of the photoreceptor.

Between physical and psychological colour characteristics there is a close relationship. The physical characteristics include wavelength, light intensity, and spectral purity; psychological characteristics are hue, brightness, and colour saturation.

There are many theories attempting to explain phenomena associated with colour vision.

The first theory of colour vision arose B.C. and has more religious basis. So, ancient Egyptian pictures indicated "emitting of particular rays by eyes that touch" the world around us. In the 5th century B.C. Empedocles [11] hypothesised about the existence of some substance flowing from the eye and from the surface of the object, as a result of their meeting there is a sensation of colour. According to Empedocles, the main colours are yellow, red, black, and white. At that very time, Democritus [12] suggested that the sensation of colour is generated by images and reflections of things "entering" the eye. According to Democritus, primary colours are red, dark green, black and white. After about 20 centuries, da Vinci [13] saw the relationship between colour and light: beauty of a colour depends on lighting, and he introduced the concept of colour contrast and determined that the white colour perception depends on surrounding colours. He considered black, yellow, green, blue, and red as the main colours.

According to modern concepts, colour perception works using visual photoreceptors on the retina – cones, where visual information is transmitted to the appropriate cerebral hemisphere due to the biochemical cycle involving rhodopsin.

At the moment there are two generally accepted theories of colour perception. The work of Marks [11] proposes to call them not "theories" but "different levels of explanation of the colour perception phenomenon"— a three-component theory of colour vision and opponent theory of colour perception processes.

1. The three-component theory of colour vision

The development of the three-component theory was made by Newton and refers to 1704 [14]. In the basis of the theory, there is a physical phenomenon called dispersion. In his experiments, Newton discovered that white light consists of a continuous spectrum of rays with different wavelengths.

In the 18th century, Newton found that any colour can be obtained by mixing the three primary colours in different proportions. The idea that any colour can be obtained by changing the intensity of three different rays is called trichromaticity.

For the first time biophysical perception of light was proposed by Lomonosov. In "the tale of the light origin that represents a new theory about colours presented in July 1, 1756" Lomonosov [15] introduced the following provisions:

- three primary colours (red, green, and yellow) and that is the minimum number of colours that in various combinations allows us to create all the colour tones;
- impact of colour on the eyes is different in character but unified by nature ("circular motion of the ether"); and
- only three zones of the spectrum are enough and necessary.

In 1802, Young [16] put forward the following theory: he assumed that each point of the retina must contain at least three structural units, sensitive to red, green and purple colours. Further development of this theory was in the works of Helmholtz [17] who suggested the existence of three types of photoreceptors with high sensitivity to blue, green, and red colour, all the three types of receptors are able to perceive other components of the visible spectrum but to a lesser extent, that is the sensitivity of receptors experiences different types of mutual overlap. The decisive experiments that confirmed the theory of Young–Helmholtz were held in 1959. Brown and Wald at Harvard University [18] and Marks et al. at the Johns Hopkins University [19] studied using an optical microscope the ability of certain cones to absorb light with different wavelength, and found only three types of cones.

It should be noted that the theory of trichromaticity found its indirect corroboration and by means of psycho-physiological methods scientists found that light sensation is a mixture of three primary colours, how selective bleaching of receptors under the action of monochromatic light influences colour vision, and also they researched colour blindness.

Thus, the colour sensation is nothing but a result of the operation of the three-component system (receptors of three types).

The difference in selectivity of cones located on the retina and responsible for colour vision is due to the presence of photopigments. Marks et al. [19] isolated photopigment from individual cones and measured the adsorption of light rays.

Spectral curve of light absorption by cone and rod pigments (P) of a human eye is shown Figure 3.1.

As follows from the results of their research (Figure 3.1) according to the dependence of absorption (in percentage terms) on the wavelength of the visible spectrum, all the cones can be divided into three types:

- cones of the first group best absorb short-wave light with a wavelength of 445 nm (cones of S-type, blue-sensitive);
- the second group absorbs medium wavelength light with a wavelength of 535 nm (M-cones sensitive to green); and
- the third group is largely absorbs light with a wavelength of 570 nm (L-cones, red-sensitive).

Research of Marks et al. [19] showed that adsorption curves are capable of mutual overlapping. White beam, in turn, equally stimulates all three cones types, causing the sensation of white.

Three-component theory received further corroboration in the works of Rushton [20] using a different approach. He proved the existence of green and red photopigment and suggested the existence of blue one.

A recent study [20] found that cones that characterise a specific photopigment differ in number and location in the central fovea and paracentral area. S-type cones are 5–10% of the total, 2/3 of them are most sensitive to long wavelength light, 1/3 is sensitive to medium wave. Cones of M- and L-type

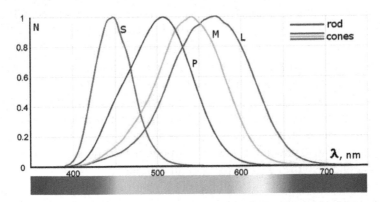

Figure 3.1   Spectral curve of light absorbtion by cone and rod pigments (P) of a human eye. S, short-wave emission; M, medium-wave emission, and L, long-wave emission.

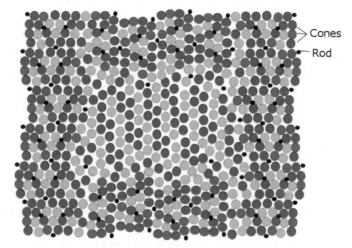

Cones
Rod

Figure 3.2 Retina receptors nearby the eye fovea.

are concentrated in the middle of the central fovea (Figure 3.2), and cones, responsible for short wavelength absorption, are on the periphery.

Cones, located near the central fovea, are thinner than peripheral cones. As the distance from the central fovea increases, the density of cones significantly reduces, area which is 3 mm farther from the central fovea has more density of the rods, when moving to a peripheral portion of the retina, the density of cones increases again. The spatial arrangement of cones determines the resolution of the eye. Since the shortest distance between the cells (the highest density) is typical for central fovea, resolution here is more, and when moving to the periphery, resolution, in the first turn the colour one, reduces [21].

2. Opponent theory of colour perception processes
Parallel to the colour theory of Young–Helmholtz another theory existed, the founder was Hering. Hering [22] interpreted the results of colour mixing based on the assumption that on the retina of the eye (or in the brain at a higher level), there are three opponent processes (or neurophysiological systems): for colour perception of red and green, yellow and blue, black and white. Thus, in addition to basic colours of trichromaticity theory, the opponent theory also presents the fourth, yellow colour. Theory of Hering states that mechanism of colour perception is such that each pair of opponent processes capable of causing only one of the two possible sensations that are included in a separate opponent process.

Since the middle of last century with the help of scientists Hurwicz and Jameson [23] the theory of the opponent processes has reached a new level. They hypothesised that receptors with a specific photopigment (S, M, or L) are associated with three pairs of neural opponent processes at higher levels of the visual system. According to them, black-and-white process is determined by transmission of light intensity, blue–yellow and red–green transmit sensations from the colour of the background, and one psycho-physiological process opposite to another in each pair of neural processes (Figure 3.3).

Thus, in accordance with the views of Hurwicz and Jameson, the information about wavelength is firstly processed in the retina by receptors that are the cones with a particular pigment (trichromaticity theory of Young–Helmholtz) and then goes to a higher level of the visual system, where three opponent processes occur (Hering's theory). In other words, the opponent properties are revealed during stimulation of trichromaticity receptors.

The theory of Hering was confirmed with the appearance of the modern research methods. Using functional magnetic resonance imaging, the existence of cells in the brain, which antagonistically react to the stimulation of red and green, yellow, and blue, was proven. They are activated by wavelengths corresponding to one end of the spectrum of visible waves, and are inhibited by the waves of the opposite end of the spectrum. Discovery of colour-opponent cells formed the basis of the two-stage theory of colour vision

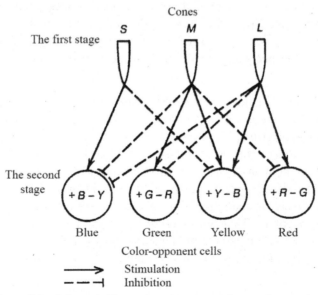

Figure 3.3   Schematic illustration of two-stage theory of perception.

(Figure 3.3), according to which the colour information is first processed by the photoreceptors of the retina in accordance with a three-component theory of colour vision (stage 1), and then by colour-opponent cells at a higher level of the visual system (stage 2).

In the first stage of colour vision, in accordance with the theory of Young–Helmholtz, but unlike the theory of trichromaticity, images are not transmitted directly to the brain. Instead, the neurons of the retina (or of higher divisions of the visual system) code a colour using the opponent signals. The output signals from all types of cones are summed (S + M + L). Then there is highlighting of the red–green and yellow–blue opponent signals which act oppositely (Figure 3.4).

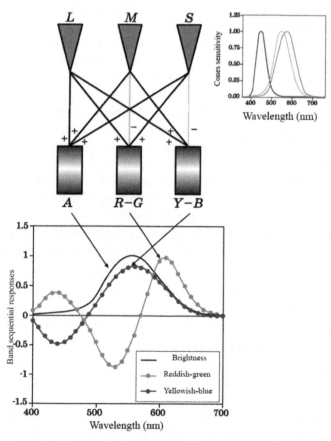

Figure 3.4 Schematic illustration of cone signals encoding to opponent colour signals in human visual system.

Thus, the two most famous theories can be combined into one called zone theory, or theory of Kries.

In 1903, Konig [24, 25] suggested in his papers that the luminosity caused by operation of the special receptor mechanism and the perception of colour is provided by at least two other receptor mechanisms.

Receptor mechanisms are composed of groups of cones with a particular band of spectral sensitivity. Theories that arise on the basis of such assumptions are called dominator-modulatory ones [26], where dominators determine the sensation of luminosity and modulators do the sensation of colour.

There was model that assumed an increase in the number of primary colours. So, "polychromatic" theory of Hartridge [27] assumed the existence of 4–5 additional receptors. At present, the theory of Hartridge was rejected due to the presence of a large number of receptors, but is widely used in polychrome seven-colours printing.

In 1955, Smirnov [28] proposed the following concept: all three types of receivers are in the same cone. In the model of the Dutch scientist, Vos and Walraven [29] there is assumed existence of three types of cones, the signals of which are divided into 2 or 3 parts, and overall the model links the theory of trichromaticity and the opponent theory of colour perception. Later this model of colour vision was described by Hubel (Nobel prize winner) [30].

Some researchers of colour vision explained the phenomenon of colour perception, attributing the colour to the properties of an eye and not as a characteristic of a light wave having a certain length. In the theory of Land [31], a colour, as it is perceived by an eye, is information "about the distribution of long and short light waves across the field of view".

Special attention is given to nonlinear two-element theory of colour vision that is a theory created by Remenko in 1975 [32]. The two-component model is based on two main principles:

- there are only two structural colour sensitive units on the retina: one-type cones and rods; and
- the process of formation of the chromaticity signals is non-linear.

Model of Remenko is largely biophysical and assumes that the brain does not participate in the processing of a colour signal. Based on the model of Remenko, there was created electronic model of an eye and a sensitive colorimeter.

In 2009, analytical review of Mark [33] was done. However, the work given in the review was not able to confirm the three-component mechanism of colour vision.

3.3 Physical and Physiological Characteristics of Colour

The perception, primarily, is determined by the light wavelength stimulating the visual system. To the human eye extreme frequencies of the visible range are quite individual [34].

We can choose three groups of colours and their respective spectral ranges:

- long-wave – red and orange;
- medium wave – yellow and green; and
- short-wave – blue, dark blue, purple.

From the point of view of the non-linear theory of colour vision, a colour is a physical parameter that determines the degree of influence and the spectral distribution of all radiation, indistinguishable for photodetector device, whose characteristics correspond to the characteristics of an eye of an average observer [32].

The colour perception in its turn is a subjective result of influence of the reflected wave in visible range having a certain wavelength on the visual system. Thus, colour is a product of the visual system. Colour vision in different organisms can vary greatly – birds, fish, amphibians, reptiles, and arthropods have highly developed colour vision [30].

All colours in the visible range of electromagnetic radiation can be divided into two groups: chromatic and achromatic. Achromatic ones are white, grey, and black colours in which the human eye distinguishes up to 300 different shades. All achromatic colours are characterised with the luminance, or lightness, i.e., the degree of proximity of this colour to white. The range of achromatic lights is almost equal to wavelengths of the entire visible range. Chromatic colours include all tones and shades of the colour spectrum. In the spectrum of the achromatic colour, relative to other wavelengths, a certain wavelength predominates.

Between colour sense and colour physical parameters, there is a close relationship. In addition to the wavelength, a colour is determined with the intensity and spectral purity. Each physical parameter has a corresponding psychological attributes such as hue, brightness, and saturation (Table 3.1).

Table 3.1 Connection between physical and physiological characteristics of colour

Physical Characteristic	Physiological Characteristic
Wavelength	Hue
Intensity	Brightness
Spectral purity	Saturation

Colour hue defines the difference of a colour of a given wavelength from grey of the same lightness. Tint (colour name) of colour hue is determined primarily by the wavelength (Table 3.2). Thus, colour differences are hue. Only achromatic colours have a hue, achromatic sensations don't have it.

The brightness of the colour depends on the intensity of physical parameter, i.e., the higher intensity the brighter colour appears. Achromatic colours (white, grey, and black) differ only in brightness. The lightness describes subjectively the brightness of the colour. From a physical point of view, the brightness determines the strength of light that is emitted or reflected from a surface unit perpendicular to the direction (the unit of luminance is Candela per meter, cd/m).

A particular interest belongs to the phenomenon of hue change as a result of changes in the intensity called the phenomenon of Bezold–Brucke [35]. Given the same intensity, some colours such as yellow seem brighter than blue, a wavelength of which is shorter than that of yellow. If you increase the intensity relative to long-wave light, for example yellow–green or yellow–red hue, it will appear not only brighter but also more "yellow".

The colour saturation is related to the physical parameter called spectral purity, and is a psychological characteristic that shows the relative amount (concentration) of the object's surface colour. Saturation characterises subjectively the intensity of the colour tone sensation. Pure spectral colours are called light with a fixed wavelength, i.e., monochromatic colour.

Table 3.2 Colours name and corresponding wavelengths

Wavelength (nm)	Corresponding Tint
380–470	Reddish–blue
470–475	Blue
475–480	Greenish–blue
480–485	Blue–green
485–495	Bluish–green
495–535	Green
535–555	Yellowish–green
555–565	Green–yellow
565–575	Greenish–yellow
575–580	Yellow
580–585	Reddish–yellow
585–595	Yellow–red
595–770	Yellowish–red

Adding light with a different wavelength or white (grey) to monochromatic light reduces spectral purity and monochromatic colour is perceived as less intense. When there are large deviations from the monochromatic one, the light loses colour and becomes grey. Thus, reducing the spectral purity of the monochromatic long-wave red light by adding white or grey colours makes it less intense and adds a pink hue.

Speaking about the colour specifications, no one can fail to mention the topic of colour mixing. There are two types of colours synthesis (blending):

- additive colour synthesis that is the process of obtaining different colours by adding (blending) the three spectral ranges – blue, green, and red;
- subtractive colour synthesis that is the process of producing colours by subtraction (withdrawal) emission from white.

Another colour characteristic having both physical and psycho-physiological significance is the contrast. Contrast is the ratio of difference between brightness of object and its background to their sum. Psychological explanation of contrast is consecutive images.

3.4 Colour Influence on the Organism

Speaking about the influence of colour on the body, one cannot fail to say about the development of colour perception in ontogenesis. Friling and Auer [36] revealed that different colours attract babies in different ways, and bright and light tones are preferable ones, black-and-white contrast holds attention too. At the age of 2 years, a child is focused more on an object shape, and colour is not a feature that characterises the object and is not taken into account in the object perception. At an early age, a child perceives colour as a characteristic of the object but the colour itself for the child is the feature of the object. Four- or five-year-old child possesses a relatively complete set of colour standards, and at the age of 5 or 6 years, there is learning of connections and relationships between colours, improving perceptions of colour leads to learning of colour tones of the spectrum. The child gets to know about the changeability of each colour by saturation, that the colours are divided into warm and cold ones, gets acquainted with soft, pastel and sharp, and contrast combinations of colours [37]. In the junior school age, the child not only recognises colours, but also correctly calls them, as well as correlates perceived quality with standard.

All this allow us to conclude that until 2 years the child has only colour preferences and colour sensory standards are formed just up to 3 years. Later, the child gets an idea about the ratio of colours, knows their names, and characteristics. In a further development colour perception is in the direction of liberation from stereotypes, subjective vision appears [37].

Colour has a significant impact on the formation of the psychophysiological status of a human. This effect is connected with activity of the autonomic nervous system (ANS), its sympathetic and parasympathetic divisions. According to Yanshin [38] "... a colour impacts on psychophysiological state of a person qualitatively and fully, including changes in blood composition, the dynamics of tissue healing, muscle tone contractions, the function of the cardiovascular system ... " Experimental work [39] showed that the colour exposure leads to changes in tonus of ANS, and the change in tone of ANS has an effect on colour vision. Thanks to the sympathetic division of ANS (SNS) sensitivity to blue–green part of the visible spectrum increases and the sensitivity to red–yellow part of the eye are reduced. Parasympathetic division of ANS (PNS) increases the sensitivity to red and yellow and reduces to blue and green. The perception of the red–yellow part of the spectrum causes activation of SNS and deactivation of PNS. Blue and green do depressing effect on SNS and activating PNS. Thus, depending on the emotional and functional status, ANS has "preferences" in colours [40].

In addition to the effects on the autonomic nervous system, colour has a direct impact on the central nervous system (CNS). Unlike ANS, which is undoubtedly influenced by a colour, the relationship between a colour and the human CNS has a more complex picture. Thanks to the certain divisions of CNS, human colour sensations are formed, and integrative activity of the CNS provides the functioning of colour perception and more complex forms of information processing. CNS lesions can lead to the loss of the human ability to perceive a colour partially or completely, the so-called colour agnosia [41].

Besides evident effect on the CNS, a colour is also necessary to maintain the tone of the CNS. The source [42] describes the cases when "colour starvation" (the colour poverty of the surrounding landscape and environment) led children to retarding of mental development.

The psychology of colour perception [43] notes various physiological characteristics of some colours. Red light activates the functional state of the organism, increases short-term muscle tension and the rhythm of breathing, and increases blood pressure. Red colour is used for the treatment of skin diseases. Yellow colour stimulates the CNS, it is used in the treatment of rheumatism, diseases of the internal organs of the gastrointestinal tract. Unlike red, green colour has a suppress effect on blood pressure, relieves stress.

Treatment by means of green colour is used for neuralgic diseases, migraine. The shades of blue are used in the treatment of inflammation of an eye.

Colour sensitivity is widely used in psycho-diagnostics. Goethe [44] formulated the basis of colour psycho-diagnostics, in accordance with which a colour can influence a person's character, integrity, communicative, and business skills of the individual. Using the methods of colouristics, the relationship between colours and personality type was determined [44]. According to Ovcharov [45], the combination of colours which are chosen by a child allows to draw a conclusion about his personality type. The conducted study by Ovcharov [45] revealed the presence of groups (named "colour") of children, distinguished by personal characteristics.

The first method of colour psycho-diagnostics was Rohrschach's "test of the ink spots" used for the diagnosis of personality qualities and traits, revealing synthetic activity and conditions of a non-directional association [46]. The method is based on the analysis of the products of creativity where a number of individual characteristics of a person are projected. The Rohrschach's test consists of standard drawings with colour or black-and-white symmetrical images, a subject is proposed to assume what each "blob" looks like. At each uncertain spot the subject sees a certain image connected with the association of his personality, and this is interpreted as a reflection of personality traits.

The most known method of the colour-diagnosis, at the moment, is a colour test of Lüscher [47], able to analyse the emotional–motivational sphere of the personality of the person, to determine the orientation of the individual to a certain functional status and to identify the more stable personality traits. Lüscher's colour test uses four general (blue, blue–green, orange–red, and light–yellow) and four secondary (purple, brown, black, and zero) colours. A subject consistently chooses the most comfortable colours and colour pairs, which are further interpreted. Analysis of the results allows identifying the subject's needs, his feelings, emotional state, and other psychological aspects of personality.

Having mental illnesses (schizophrenia, manic-depressive syndrome, and acute phase of neurosis) the attitude to colour changes and that might be one of the methods of determining the early stages of these diseases [48].

Thus, a colour as a psychological and psycho-physiological parameter is associated with psychosocial characteristics of personality: emotions, character, and thinking.

Undoubtedly, the indicators of perception of different colours differ in people with diseases of the visual system (glaucoma, cataract, etc.). In this case, colour methods, including the method of determining the critical frequency of flicker fusion, allow determining the state of the visual analyser and progress of the disease.

3.5 Basics of CFFF Method

Critical flicker fusion frequency method in psychophysiology based on the perception of the number of flashes per time unit. Under CFFF, we understood that the maximum frequency of the interruptions of the light per second at which an intermittent colour starts to look like blinking, but gives the impression of a smooth, with the same not flashing brightness [49].

Critical flicker fusion frequency as a characteristic depends on many factors: the size of the test field and the place of its projection on the retina, and light intensity, light spectral composition, duration and depth of modulation of the light stimuli, and their quantity in case of multiple presentation [4].

The advantage of CFFF method is its independence on visual acuity and refraction, CFFF characterises the functional state of the visual analyser in general [50].

In accordance with the Ferry–Porter's law [51], CFFF is proportional to the logarithm of the intensity of stimulating light:

$$n = a \lg I + b, \tag{3.1}$$

where n, the flicker fusion frequency; I, the intensity of the light stimulus, and a and b, constants determining the spectral characteristic of the wavelength of stimulating light. The Ferry–Porter's law, according to Golubtsov [50], is true for medium brightness, for $a < 4$, and with increasing light stimulus intensity CFFF rate decreases, if the total area of the stimulus and brightness increases.

At the moment the effect of the preliminary adaptation test, measurement conditions (the surrounding background, the exposure of the peripheral areas of the retina, etc.) and incidental stimuli (auditory, gustatory, and olfactory) on CFFF is revealed, and they can affect CFFF in both directions.

It is revealed that the rate of CFFF depends not only on photochemical processes on the retina, but also on the state of neurons at higher organisation levels of the visual perception process [49].

It is also shown that the rate of CFFF almost linearly increases with the size of the projection of the stimulus on the retina (Granit–Harper's law [52]). For the central regions of the retina when the law is true when the size of the stimulating zone is up to $45°$.

Critical flicker fusion frequency registration depends not only on measurement methodology but also on the physiological state of a person [49].

The normal CFFF rate for adults and children is 41–45 Hz. Krasnoperova [53] revealed that these indicators are characteristic only for the macular zone of the retina and only if the stimulus is presented centrally. A number of researchers [29, 30] found that for the central retina zone (5°) CFFF rate is from 40 to 45 Hz; for the paracentral zone (10°–20°) it increases up to 55 Hz, for the periphery it is reduced to 35–40 Hz. Other authors [49] claim that the rate of CFFF in peripheral retina area is 60 Hz. This difference is primarily due to the lack of a standard method and equipment. The traditional version of CFFF assumes that the stimulus is perceived mainly by macular area of the retina. At the moment, it is found that having presentation of stimulus in the angular range 10°–55°, the CFFF rate is proportional to the logarithm of the angular size of the view field and increases towards the peripheral retinal area by 10–15 Hz.

When presentation is central, normal CFFF for green stimulus is a few Hz higher than for the red one [50]. This is due to the fact that in the region of the central fovea larger amount of red-sensitive cones locates, and in paracentral region there are mostly green cones. The difference between the data on stimulation by green and red light is from 3 to 4 Hz [54]. This difference is true for all age groups except the eldest, and may serve as indicate of CFFF norm for monochromatic stimuli of red and green.

A number of researchers note the dependence of test results on fatigue of the visual analyser, in particular, after the text reading from electronic devices [4]. The CFFF dependence on the degree of fatigue is due to the fact that in the development of fatigue caused by physical or mental work, the main role belongs to the central nervous system. Fatigue of the body is the process connected with the central cortical leading element representing cortical protective reaction; by physiological mechanism, it is the reducing of efficiency of cortical cells themselves due to their protective inhibition. CFFF rate is determined including higher divisions of the visual analyser, since the central visual neuron and the visual cortex are the most inert of the visual system what leads to CFFF decreasing because of fatigue of the body that causes the reduction of cortical cells efficiency. This fact allows controlling of the functional state of the organism and the degree of fatigue according to the change of CFFF [55].

Flickering light affects not only the retina but also the centres responsible for eye movements and the passages connecting these centres with the eye muscles [49]. Thus, the flickering light of a particular wavelength and with a certain frequency can serve as a therapeutic method. Students of Kravkov [56] determined the positive effect of green stimuli for high eye pressure. To a certain extent the effect of flickering colours can ease the development of pathological processes and start the process of recovery of visual perception neuronal mechanisms activity.

3.6 Experiment Methodology

Any physiological studies involve several stages. The first step involves the selection of appropriate research subjects. The second stage does directly the experiment divided into two parts. In the first part, the values of critical frequency of flicker fusion for the red, green, blue, and white stimuli and the frequencies for better perception were determined. The second part contained CFFF determination after reading small text on laptop LCD display. In the third stage of the research, there were statistical procession and analysing the experimental data.

The subjects were young men and women from 20 to 24 years old in equal number, and the total number of subjects was 42. In the process of analysing the results from the sampling three subjects (two boys and one girl) were excluded due to the presence of colour vision abnormalities, and thus the total number of subjects was 39, 19 of which were boys.

In the process of CFFF, rate determining the main factor passing to research is the lack of nervous system diseases (epilepsy, multiple sclerosis, or other) and that of visual analyser (retinal detachment, etc.). Therefore, the first phase of studies was the selection of subjects with the absence of such diseases. At the same time, a number of authors [45, 48] state that visual acuity does not affect the value of the CFFF rate.

In the colour perception studies, an important factor is the absence of anomalies of colour vision, such as colour blindness, monochromatism. Even protanopes and deuteranopes cannot be included in the total sampling during results processing.

At the first stage of the research that represents testing the subject, one conducted the selection of persons eligible for the study: the subjects were surveyed by means of oral questionnaire to reveal the presence of diseases of the nervous system and visual analyser, including diseases in previous generations. The study allowed persons with no such diseases.

Next, the subject looked through polychromatic tables (Rabkin's tables [57]) to detect anomalies of colour vision. Only trichromates were passed to the study. It should be noted that among the nearly 50 subjects only 2 youths were protanope and deuteranope.

One of the methods used in the study was a method of determining values of the CFFF rate. Under the critical frequency, we understand that the minimum frequency of colour breaks per a second whereby intermittent colour ceases to be perceived as flashing, but gives the impression of a smooth one being of not changing brightness [49]. As noted above, the rate of CFFF depends on many factors: the projection of the stimulus on the retina, the intensity and size of the light source of flickers, spectral composition, duration and modulation of the stimuli, as well as from external factors and irritants. Another problem is the dependence of the CFFF on a functional physiological state of the subject [50].

To measure CFFF we made a choice of CFFF glasses.

3.7 Devices Comparison

Let's consider the existing devices for determining CFFF:

1. Teterina's device [58] for the human body functional systems correction is a device for correction of the functional systems of the human body consisting of glasses with two light-isolated oculars, each of them has a light-diffusing reflector having inside rigidly fixed light emission source, driving oscillator, the light flicker frequency controller, and the power supply. The latter one is distinguished by having a light brightness controller, connected in series control unit; the modulated pulses generator and modulator having inputs connected, respectively, to the outputs of the light flicker frequency controller and outputs connected to the respective inputs of the light brightness regulator; two switches, each has its input connected to the light brightness controller and output to the light source being a set of light-emitting diode (LED) emitters of different light wavelength; the control unit the outputs of which are connected to the control inputs controls the light flicker frequency, light brightness control, and switches.

Disadvantages of the device:

- This is a completed device, unchangeable for the specific needs, all the parameters are strictly defined, modifying of their values is not possible;
- It has only 4 default brightness modes; and
- A small range of oscillations with fixed increments.

2. The device for light influence on the human body. The invention relates to medical devices. It contains glasses—photic stimulators with coloured

light emitters mounted in lightproof mask-spectacle frame with light-isolated oculars and connected to the electronic control unit which includes a phototherapy procedures execution and installation control unit (being microprocessor and emission control unit and having its inputs connected to the outputs of the microprocessor, and outputs—to the emitters of glasses-photic stimulator). The device also contains autonomously reprogrammable memory device (connected with phototherapy procedures execution and installation control unit), power supply, keyboard, procedures LED light, and system bus coupled to the microprocessor. The device is also provided with additional microprocessors (built-in electronic control unit on a multiprocessor architecture, and connected by means of additional microprocessors via the system bus with the microprocessor of phototherapy procedures execution and installation control unit), each of which is provided with its own interface with outputs for connecting the sensor of psychophysical state of the patient and sensor of the patient eye status check, respectively. Use of this device allows extending the functionality of the devices for light exposure on the human body [59].

Disadvantages of the device:

- no possibility of combining signal form and brightness, including for each individual light source since to change opportunities connection to a computer is required; and
- lack of autonomous work in the absence of an industrial network.

3. Chromotherapeutic LED device contains put-on-the-head case with the light source in form of LEDs, power supply and control unit. The LEDs are divided into groups with red, blue or green emission spectra and white light illumination. The control unit contains a microprocessor system with the groups of LEDs switch, LEDs brightness regulator, the general LEDs brightness regulator, unit for connection to a personal computer with the software for colour combinations, pulse duration and brightness generating. In the device of the first embodiment the case was designed as a visor located above the face. LEDs were arranged at the bottom of the visor base fastened to the belt. According to the second embodiment, the case was created in the form of spectacles with lenses made of transparent plastic and frame. Frame consisted of two sidewalls and plastic was located along an arc between the sidewalls. LEDs were placed in the sidewalls at both sides of end parts of plastic so that the light flux from them was directed inward and to the centre of the plastic. To form

the movable coloured light spots on the surface, the latter was made of transparent material and was bent along an arc. The light sources were disposed between the layers of material at the ends on both sides in the surface. Their luminous flux was directed towards each other and the amount of light and colour of each light source is controlled according to various laws [60].

4. IMEA ADR III critical flicker fusion frequency measuring device [61].

Disadvantages of the device:

• the device has a complex structure consisting of a set of electronic components;
• its usage is possible only with the participation of experienced specialists; and
• for prophylactic chromotherapy a special room is needed.

3.8 LED Unit

Light-emitting diode unit solves the following tasks: evaluation of the person functional state according to indicators of fatigue of the visual analyser and sustainability of thresholds of light flashes distinguishing; revealing of risk groups among PC users (especially children) according to the increase in fatigue when working with the computer, professional selection, and preventive examinations of employees whose work is related to the long eyestrain as well as diagnosis of the number of eye diseases and other diseases (in particular, diabetes).

This objective is attained with the light module sending separate flashes of light (when flashing frequency increases above a certain level there are no longer ability to distinguish them, that is they merge and are perceived as a continuous light).

The device is hardware and software system consisting of:

1. Blackout glasses with built-in LED emitters and matt diffusers.
2. Control unit (CU) that is a micro-controller control circuit and used to receive control commands from a personal computer (PC) or from any other device that supports used interface and protocol, as well as to transfer the information you need to the PC.
3. The personal computer (PC) – control terminal is used for controlling the CU and providing interface for interaction with the user.

Light-emitting glasses (Figure 3.5) represent a light-isolated glasses (1) having three-point LED of red, blue, and green colours on the distance of 25–30 mm

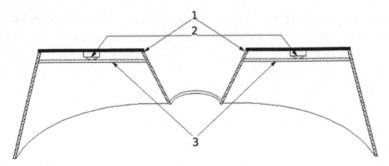

Figure 3.5   Light-emitting glasses.

from pupils (2). Due to the fact that the LEDs crystals form a triangle with a side length of not more than 1.5 mm and due to the use of matt diffuser (3) the glow of all three LEDs is perceived as a point that allows synthesising any colour being perceived by the human eye by adjusting the brightness of each LED.

The control unit (CU) is the main unit of the system. CU is a micro-controller system with USB interface for PC connection. Power of system including LEDs is from USB PC interface so that the device does not require additional power source.

Figure 3.6 is a schematic diagram of the LED unit.

Using asynchronous pseudo-multitask work mode allows building flexible control algorithms and changing any parameters of the experiment directly from the PC side during the experiment. This mode is provided by the use of two microcontroller hardware interrupts: when overflowing of counter timer (this timer is responsible for the internal synchronisation of processes

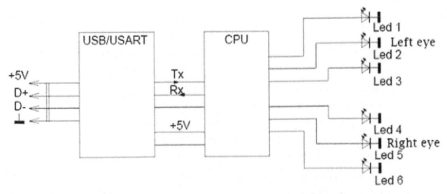

Figure 3.6   Schematic illustration of the LED unit.

and provides the stability and accuracy of the specified parameters retention) and interruption when data arrives via Universal Synchronous-Asynchronous Receiver/Transmitter (USART) interface of microcontroller (which provides immediate response to all incoming data).

This approach provides rational use of microcontroller resources, fast operation, instant response to incoming commands, stability and accuracy of the specified mode retention, and overall flexibility and universality of the system.

The advantage of this LED unit of model is the ability to configure the device for the specific purposes. It is possible to set them during the session as well as to use a predetermined program. Settings include: the flicker frequency of $0 \ldots 100$ Hz, with increment -0.1 Hz independently from each other, control of each colour independently for each constituting colour, control of brightness of each colour is in the range from 0 to 100% for each colour. The light source is LED and that increases lifespan of consumables and lowers power consumption. National Instruments equipment was used while given device developing and adjustment [62].

3.9 Software

The program unit solves the following tasks: evaluation of the person functional state according to indicators of fatigue of the visual analyser and sustainability of thresholds of light flashes distinguishing; revealing of risk groups among PC users (especially children) according to the increase in fatigue when working with a computer, professional selection, and preventive examinations of employees whose work is related to the long eyestrain and diagnosis of number of eye diseases and other ones (in particular diabetes).

CU includes software allowing to:

- accept commands from the PC;
- transmit data to the PC; and
- in asynchronous pseudo-multitask mode perform the following subprograms: independent blinking with a predetermined frequency of any combination of LEDs at a frequency from 0 to 100 Hz with increment of not less than 0.01 Hz, independent adjustment of the brightness of each LED in the range from 0 to 100% with 1% increment, independent enabling or disabling any LED;
- storage of configuration data (calibration constants) in the electrically erasable programmable read-only memory (EEPROM) of the microcontroller, which allows the calibration of the instrument from a PC without making any changes to the schematic diagram and managing program.

The algorithm of the program is as follows (Figure 3.7):

- when you run the managing program of the microcontroller the synchronised with the crystal oscillator timer counter starts;
- loading configuration data from the EEPROM into the corresponding configuration flags;
- initial values are recorded into the control flags (responsible for the frequency of blinking, brightness of each LED, flags are enabled/disabled for each LED, etc.);
- when interrupts of timer counter appear, respective LEDs get on or off according to the current state of the internal counters and flags;
- timer switches again to initial state and standby mode while waiting for the next interrupt due to the overflow;
- when triggered by interrupt via interface USART because of data entry data go through a filter-former of commands;
- at the end of the command formation, command parser runs what makes command's analysis along with input data validation conforming to designated boundaries for them; if the data conform its boundaries, flag of command changes to a new value; and
- next, the system goes into standby mode waiting for the next interruption due to the overflowing of the counter timer or due to data arrival via interface USART.

Control terminal (CT) provides interface for interaction with the user and control of CU. It is software system that provides: CU management, processing of experimental data, formation of documentation of the experiment, and the database of patients studied, which also allows statistical data processing. Software package is implemented in the C++ programming language using the framework Qt, providing cross-platform property and allowing to work on different operating systems (including many UNIX systems and, respectively, Linux-operating systems) and on platforms with different architectures of managing processor (IBM compatible, ARM, and so on). The software has a user-friendly interface with protection against incorrect user actions that provides full control over the measurement process and eliminates the possibility of breaking the system or its malfunction due to incorrect actions.

Expected result from the use of this utility model is to identify the CFFF. In ophthalmic practice, CFFF is a reliable criterion for the diagnosis of a number of eye diseases and other ones (such as glaucoma, diabetes) during professional selection and preventive examinations of workers whose professional activity is associated with prolonged eyestrain.

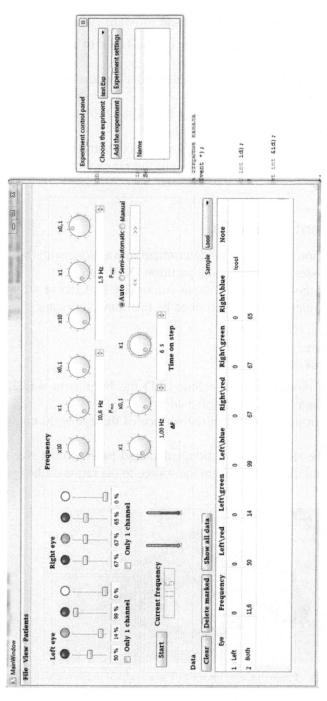

Figure 3.7 Main window of the program.

The main characteristics of hardware and software system:

1. Polychromatic LED device formed on the basis of the light-isolated glasses.
2. Installed LEDs can implement a visible colour spectrum by mixing three primary colours.
3. It can set flicker frequency in range from 0 to 100 Hz with increment −0.1 Hz independently from each other.
4. Colour management independently for each constituting colour.
5. Brightness control for each colour is in the range from 0 to 100% for each one.

3.10 Experiment Performing

Critical flicker fusion frequency rate determination was done with the help of original CFFF glasses described in the partition IX (Figure 3.8).

The source of the stimulus was three-colour (RGB) LED of 4 mm size. The dominant wavelength was determined by the monochromator:

- for red source: 630 ± 15 nm;
- for green source: 525 ± 15 nm;
- for blue source: 470 ± 10 nm.

For all operation modes of three-colour LED the brightness was 30% of nominal, that is, on average, from 4 to 5 cd/m^2.

The distance from the source to the surface of the eye is 20 mm on the average.

The area of the stimulus projection, calculated based on the size of the stimulus source and the distance from the source to the retina of the eye, was

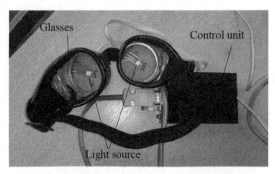

Figure 3.8 Photo of the original CFFF glasses used in the research.

about 60°. Thus, stimulating area captured the central, paracentral, and part peripheral area of the retina.

Critical flicker fusion frequency glasses have a connection to a personal computer via USB, the control is performed using specialised software "eyeLight" that lets you specify the spectral composition, intensity, depth and duration of the stimulus (Figure 3.7).

The studies were conducted sequentially (firstly on one then on the other eye) in the direction of increasing frequency at a speed of 1 Hz/s for white, red, green, and blue colours for each eye separately. The subject was asked to mark the moment of flicker fusion, at the same time the experimenter using the software "EyeLight" fixed this point and the frequency. Doubtful experimental values of CFFF rate were put to check by repeated presentation of the stimulus several times, the analysis was made for the results matching at least two times.

The experiment was repeated for one eye (usually for the right one) in 10 min after the subject had read a text on a small LCD located at a distance of the best vision with the aim of identifying visual analyser fatigue influence on CFFF rate.

Determination of the value of the object colour tone matching the background at different playback speed. The next method which has a number of disadvantages associated with different reaction speed of the subjects, but nevertheless allowing to estimate influence of frequency change (frames/second) on the perception of red, green, and blue colour in the computer program "Colour Intensity" (Figure 3.9).

In the window of the tested person the saturation of the colour tint of the object was changed with certain rate of frames per second compared to the background colour, the working speed of playback was selected 20, 40, and

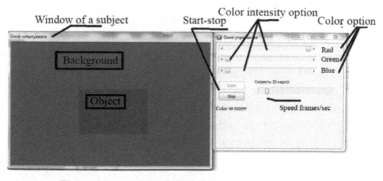

Figure 3.9 Interface of "Colour Intensity" software.

60 frames/s. At the moment when the subject ceased to perceive the differences between colours of the object and the background, the program operation was suspended: the control window displayed the value of matching of the object colour with the background colour in percentage. In most cases, the experiment was repeated several times, and data that matched at least in two repetitions of the experiment were recorded.

The control window allows you specifying the colour intensity (red, blue, and green in various ways) and the rate of change of colour saturation of the object with respect to the background colour.

The correlation between the experimental results obtained by both experimental methods is of special interest.

3.10.1 Mathematical and Statistical Processing

The last stage involves the statistical processing of the experimental results and the visualisation of the data. Statistical processing of the research results was carried out using a table editor Microsoft Excel Microsoft Office and specialised software for statistical data processing IBM SPSS Statistics v.20. Statistical processing of the results consisted of an analysis of the distribution of features and their numerical characteristics (average values, error of the mean, and standard deviations). To assess the reliability of the differences there were Student t-test and Fisher's test used. Correlation analysis involved the correlation coefficient of Spearman. When comparing samples one used the criterion of Kruskal–Wallace. Differences were considered significant when the value of the significance level was $p < 0.05$.

3.11 Study Results

1. CFFF rate determination

The main difference between the original CFFF glasses used in this chapter and factory equipment is a shorter distance from the stimulus source of light to the retina: serial devices have an average distance of 70–90 mm while the device used has 20–40 mm. Because of such stimulus source position with respect to the retina CFFF rate may be different from those obtained using factory equipment. In the first place such difference may be due to the size of the stimulus projection on the retina: the greater distance to the source of flickers the smaller size of the projection (Figure 3.10).

Critical flicker fusion frequency rate dependence on the stimulus projection size on the retina is related to the distribution of cones with a certain

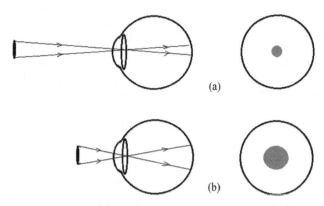

Figure 3.10 Stimulus projection size on retina in case of (a) large distance between source and retina, (b) small distance between source and retina.

type of photopigments on the retina. The distribution of various types of cones on the retina is uneven and, therefore, CFFF rate for different colours is different.

Normal CFFF rate of adults and children is 41–45 Hz at the stimulus position in centre and only for macular area of the retina. There are different opinions of various authors about CFFF rate changing with increasing of stimulus field projection: earlier CFFF rate for the peripheral area of the retina was considered to be decreasing to 35–40 Hz, more recent studies revealed that CFFF rate for peripheral retina is 60 Hz. It is now determined that when locating the light stimulus in the angular range from 10° to 55° the CFFF rate is proportional to the logarithm of the view field angular size and increases towards the periphery by 10–15 Hz [4].

When locating at the centre CFFF for green stimulus normally is few Hz higher than for the red one. This is due to the fact that in the central fovea area there are lots of red-sensitive cones, in the paracentral area there are primarily green ones. Thus, the difference between the data on stimulation by green and red is 3–4 Hz. This difference is reliable for all age groups except older one and may indicate the normal CFFF rate for the monochromatic stimuli of red and green colour.

In our study (Table 3.3), the maximum values of CFFF rate were observed when showing green stimuli and its values were 47.6 ± 1.1 Hz. Then that procedure was repeated for red stimulus and received values were 45.4 ± 1.1 Hz. The lowest CFFF rate was when showing the blue stimulus, and it was 43.4 ± 0.5 Hz. When showing polychromatic white colour stimulus, the

Table 3.3   Results of CFFF

Eye	Stimulus Colour	CFFF Value (Hz)	SEM	Standard Deviation (SD)
Left	White	47.45	0.52	3.22
	Red	44.37	0.63	3.87
	Green	46.53	0.63	3.86
	Blue	42.82	0.86	5.31
Right	White	48.08	0.6	3.67
	Red	46.63	0.48	3
	Green	48.63	0.51	3.13
	Blue	43.89	0.9	5.4

average CFFF rate was 47.8 ± 0.3 Hz that was maximum one. The data support the hypothesis about showed stimuli colour dependence on the number and location of the cone cells of a certain type which receive the radiation of the visible spectrum of different wavelengths on the retina.

The average CFFF rates differ slightly from those given in the literature for the paracentral area of the retina [4]. The paracentral area of the retina is rich in cones with photopigments being responsible for the perception of the medium wave range of the spectrum, so the CFFF rate has a maximum value exactly when showing green stimuli. In the central macular area of the retina, it is prevalence of red-sensitive cones but their number in the projection of the stimulus on the retina is marginally less. The lowest number of cone cells that are responsible for the perception of short-wave radiation (5–10% of the total number of cones) explains a minimum ratio of CFFF when showing blue stimulus.

Given correlation analysis by CFFF rate values (Figure 3.11) for different colours allows asserting that conducted research corresponds to colour vision theories.

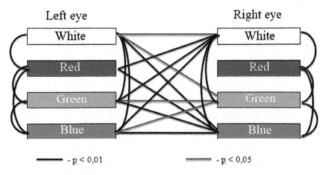

Figure 3.11   CFFF rate correlation for both eyes.

In some cases the lack of correlation for CFFF rate between the right and left eye when presenting a stimulus of the same colour is due to the fact that 20% of the subjects had different visual acuity for each eye. With excluding these data from the sampling, correlation between the CFFF rate of one colour stimulus for the left and right eyes was shown.

Thus, studies conducted also represent a challenge to the assumption that CFFF rate does not depend on the visual acuity [48–50]. As we know, abnormality of visual acuity leads to eye optical system focal length displacement to the front of or behind the retina (Figure 3.12), as a consequence the projection field of the stimulus increases or decreases.

In its turn, the studies presented CFFF rate dependence on the projection field associated with different quantity of cons involved into colour recognition. In that case projection, field size is defined not so much by eye vision

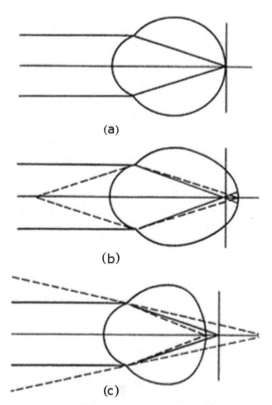

(a)

(b)

(c)

Figure 3.12 Ray course with different eye refraction types (a) normal vision acuity; (b) myopia; and (c) hypermetropia.

Table 3.4   CFFF rate difference for different colours of stimulus ($D < 0.01$)

Stimulus	CFFF Rates Difference (M ± m), (Hz)		
	Red–Green	Green–Blue	Red–Blue
Left	2.16 ± 0.39	3.71 ± 0.60	1.55 ± 0.48
Right	2.00 ± 0.34	4.74 ± 0.81	2.74 ± 0.89

acuity abnormality type (myopia or hypermetropia) as by the degree of the abnormality (weak or strong).

The difference of CFFF rates when showing red and green stimuli can be evidence of the CFFF glasses proper operation (Table 3.4). During our research the difference in CFFF rates when showing red and green stimuli was 2.08 ± 0.08 Hz. The diversity in the difference rate in 3–4 Hz obtained by the authors [4] is also associated with a large projection field of the source when showing the stimuli.

Some researchers claim that the figures for the CFFF of red and blue stimulus are almost equal [50, 54]. Our research casts doubt on such statements, because we noted the difference between CFFF rates:

- the average value for blue flashes is less than that for red flashes by 2.15 ± 0.60 Hz; and
- the average value of CFFF for green flashes is greater than when showing blue ones by 4.2 ± 0.8 Hz.

That can be explained probably with constructional features of used CFFF-glasses associated with the light source stimulus projection going beyond the bounds of the macular central area of the retina where the distribution of cones with pigments of certain type is different from that of central region.

The value of the CFFF rate is known to change with age [48–50]: after reaching the age of 40, CFFF begins to decline by 0.2 Hz/year on the average, reaching values of 38 Hz up to 70 years. At the same time, it is believed that there is no difference in CFFF rate between men and women. Studies [63] found that when using the "EyeTracking" method there are certain differences in the visual perception of dynamic monochrome black-and-white objects.

The analysis of gender peculiarities of perception of the colour characteristics was carried out for values of the CFFF rate values, averaged over right and left eyes. Our data show the existence of differences in the CFFF value for subjects of different genders (Figure 3.13). So, when presenting stimuli of white colour, difference in the value of the rate is 2.4 ± 0.2 Hz, 1.9 ± 0.1 Hz for monochromatic red and 5.8 ± 0.6 Hz for monochromatic blue.

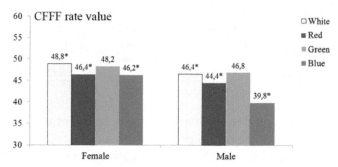

Figure 3.13 CFFF rate for stimuli of different colours for the male and female subjects.
*Significance level of differences between gender groups $p < 0.01$.

The female tested had the CFFF rate value reliably higher for all used colours of light stimuli.

The greatest difference among CFFF rates of the subjects of different genders was identified for blue flickers. Possible cause for large differences in CFFF rates when blue stimulus presented for girls and boys may be the fact that there were more girls with blue colour cornea than boys. As it is known [34], blue cornea largely lets short-wave radiation pass compared with the brown cornea. The number of photons of short-wave radiation passing through the blue cornea will be absorbed to a lesser extent and the CFFF rate in this case will be bigger.

A correlation analysis (Figure 3.14) confirms the presence of ties within groups by gender of subjects and lack of connections between groups, which

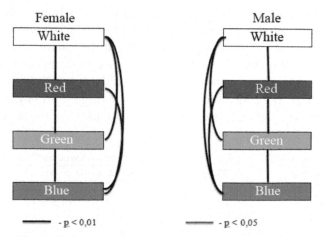

Figure 3.14 CFFF rates correlation for males and females.

Table 3.5   CFFF rate difference for different colours of stimulus ($D < 0.01$)

Gender	CFFF Rates Difference ($M \pm m$), Hz		
	Red–Green	Green–Blue	Red–Blue
Female ($n = 21$)	1.86 ± 0.28	2.00 ± 0.48	
Male ($n = 17$)	2.35 ± 0.36	6.97 ± 0.74	4.62 ± 0.65

allows stating about the gender peculiarities of perception of colour stimuli and the CFFF rate dependence on the gender of the subjects.

The difference in the CFFF rates when presenting stimuli with different colours also differs depending on the gender of the subject (Table 3.5). Male subjects had the difference in CFFF rate bigger than that of girls. Thus, female subjects had more stable differences between CFFF rates.

2. Determination of match of an object colour with the background one having different speed of playback

Analysis of data on the percent matching of an object colour and background one at different playback speeds was implemented using software "Colour intensity" that allowed determining the influence of the object colour compared to the background colour on the perceived playback speed (frames/second). The result of deviations is presented in percentage where 100% is when object colour matches background one. The method used the following selected playback rates: 20, 40 and 60 frames/second. It should be noted that significant differences in rates for males and females were not found, so there are deviation in values averaged over the gender of the subjects (Table 3.6).

Values of object colour tints matching the background colour tints (Table 3.6) are correlated with values of the CFFF rate (Table 3.3). In this case, for the background-object of blue colour there is maximum deviation that may indicate a worse perception of changes in the intensity and colour tint for short-wave sources. Deviations are not so significant for stimuli of medium-wave and long-wave areas. This is probably due to the dependence of the absorption of short, medium and long waves (Figure 3.1) by the retina on the number of cones of different types. Another reason for the large values

Table 3.6   Values of matching of colour tints of object and background, % ($D < 0.05$)

Colour ($n = 39$)	Playback Speeds (Frames/s) ($M \pm m$)		
	20	40	60
Red	99.09 ± 0.08	98.23 ± 0.23	97.18 ± 0.14
Green	99.10 ± 0.13	97.93 ± 0.24	96.85 ± 0.14
Blue	98.77 ± 0.14	97.76 ± 0.13	96.58 ± 0.16

of the deviation for blue colour in given method and smaller values of CFFF rate for blue colour stimulus is the colour of the cornea: the brown colour cornea absorbs more intensively short-wave region of the visible wavelengths spectrum of electromagnetic radiation.

Correlations (Figure 3.15) show the relationship between the colours of the background and that of object within the group by deviation playback speed of the object colour compared to background one. At the same time, the greatest number of correlations are observed between the groups by the playback speed of deviations of the colour tint equal to 40 and 60 frames/s. This fact suggests that frequency of better perception of the playback speed is close to a value of 40 frames/s.

Correlation analysis of links between the rates of CFFF for stimuli of red, green, and blue colour and the value of object and background colour matching with playback at different speeds for red, green, and blue colours objects indicates the predominance of connections between the CFFF rate and the value of matches during playback for groups of colours. That is, the average CFFF rate for all subjects, for example, when presenting green stimulus, is associated (when the coefficient of significance $p < 0.05$) with the matching of object colour with the background colour at their green colour tint regardless of playback speed of change of the object–background colour.

Thus, CFFF rate correlates only with the results of the match of object and background colours and it is not a function of playback speed of object–background colours.

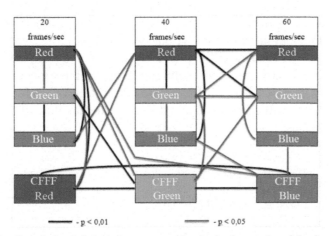

Figure 3.15 Correlation of deviation rates with playback speed and background–object colour tint.

3. Estimation of visual analyser fatigue

In the work of Morozova and Novikova [8] using the method of CFFF the visual analyser fatigue value was determined. It was found that when reading text on LCD the value of CFFF rate compared with the background decreases by 3.0 ± 0.2 Hz. Given work also studied the visual analyser fatigue during the presentation of stimuli of white and monochromatic red, green, and blue colours for the right eye during the background measurement (Table 3.3) and the measurements after reading small text at a distance of best vision on the display of the laptop for 10 min (Table 3.7).

It is worth noting that correlation analysis between the CFFF rates when presenting the stimuli after reading the text on the laptop display didn't reveal any correlation between the groups chosen by gender. That is, the relationship between the rates (at the significance level $p < 0.01$) exists only within the group of the same gender between the values of CFFF when presenting the stimuli with different colours.

Comparing CFFF rates during the background measurements with the measurements after reading the text, you can draw some conclusions:

- after reading the CFFF rate significantly decreased by 2.20 ± 0.06 Hz when presenting white stimulus, by 2.26 ± 0.08 Hz when presenting red one, by 3.32 ± 0.21 Hz when presenting green one (for stimulus of blue colour rate changing is not reliable; Figure 3.16);
- as a result, there were found statistically significant differences when comparing CFFF in the background state and after stress, and in comparison with the results recorded in a background state; after stress, the CFFF rate reliably decreased, it does not correspond to normative values what speaks about fatigue of the visual system;
- the difference between the CFFF rates after the text reading is statistically significant ($p < 0.01$) only for the difference in CFFF for green–red, and it equals to 2.1 ± 0.5 Hz, and for females this difference is bigger; and for red–blue, which equals to 1.9 ± 0.3 Hz, whereas this difference is bigger for males.

Table 3.7   CFFF rates for a right eye after reading small text for 10 min ($M \pm m$)

Rate	White	Red	Green	Blue
All ($n = 39$)	45.82 ± 0.6**	44.29 ± 0.6	45.11 ± 0.8*	44.18 ± 0.8
Female ($n = 21$)	46.86 ± 0.8**	44.81 ± 0.8**	47.57 ± 0.7*	46.43 ± 1.0*
Male ($n = 19$)	44.53 ± 0.6*	43.65 ± 0.8	42.1 ± 1.2*	41.4 ± 1.2

*$p < 0.01$, **$p < 0.05$.

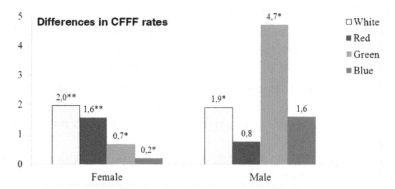

Figure 3.16 CFFF rates difference for various stimuli colours of the male and female subjects.

*Significance level of differences between gernder groups $p < 0.01$;
**Significance level of differences between gernder groups $p < 0.05$.

The system of the human visual analyser is not enough adapted to read information on LCD screens. The image on the display monitor differs significantly from natural objects – the image on the screen is self-illuminating and is not reflected one, and has less contrast. The icons on the screen are represented by a number of discrete points—the pixels—which do not have clear boundaries. These reasons explain the decrease of accommodative focus that leads to the formation of the lag of accommodation. Accommodative muscle is no longer able to keep a focus in the right position and often the focus is behind or in front of the screen in the stationary point of accommodation. Eye constantly switches from stationary point of accommodation to the point of focus on screen, and that leads to visual fatigue. When working with LCD displays, there is a need to identify small objects, as a result, the main burden falls on the macular area of the retina, which determines the rate of CFFF.

3.12 Conclusion

The chapter presents the theory of the colour vision and identifies physical and psycho-physiological characteristics of a colour, especially the perception of a colour.

The analysis of the glasses available for the CFFF studying was conducted. Based on the analysis a new model of glasses was created.

Expected result from the use of this utility model is to identify the CFFF. In ophthalmic practice CFFF is a reliable criterion for the diagnosis of number of eye diseases and other ones (such as glaucoma and diabetes), during professional selection and preventive examinations of workers whose professional activity is associated with prolonged eyestrain.

The main characteristics of hardware and software system:

1. Polychromatic LED device was formed on the basis of light-isolated glasses.
2. Installed LEDs can implement a visible colour spectrum by mixing three primary colours.
3. It can set flicker frequency in range from 0 . . . 100 Hz, with increment −0.1 Hz independently from each other.
4. Colour management independently for each constituting colour.
5. Brightness control for each colour is in the range from 0 to 100% for each one.

The experiment was conducted. The results of the study of psycho-physiological features of perception of colour perception intensity allow drawing the following conclusions.

1. CFFF rate received is different from those cited in the literature. This difference is due to design features used in the CFFF-glasses work: the smaller the distance from the stimulus source to the eyes, as a consequence, the larger projection field of the stimulus on the retina.
2. CFFF depends on the colour of the stimulus. The direction of the CFFF rate decreasing the colours of the stimuli is as follows: white → green → red → blue. This reduction in CFFF along with decrease in the colour wavelength of the source is associated with the number and arrangement of cones responsible for absorption of radiation of a certain frequency.
3. The difference between the CFFF rates is reliable for the following pairs of colours: red–green and green–blue. This is a consequence of the large projection field of the stimulus, including central, paracentral, and part of the peripheral region of the retina.
4. The gender features of visual perception correspond to the CFFF rate: depending on the wavelength of the stimulus source the CFFF rate of females is by 2–5 Hz greater than that of males.
5. The value of matching of the object colour hue with the background colour is determined largely by the rate of colours change of the object–background. For blue colour, object–background matching is smaller than for red and green. The colour of the stimulus during definition of the CFFF rate correlates with the object–background colour and does

not depend on the rate change of the object–background colours tints saturation.

6. After reading the text on a laptop LCD display the CFFF rate significantly decreased for all wavelengths of the stimulus source and may serve as an indication of visual analyser fatigue at the level of the retina as well as at higher neuronal levels of the brain.

In conclusion, it is worth noting that the CFFF can serve not only as the indicator of normal visual perception, but also as one of pathology and fatigue of the visual analyser.

Acknowledgements

The authors would like to thank Anatoly Sachenko, Anna Monovskaya, Yury Kolokolov and the authors contributing to this book for their professional advice, valuable feedback on a preliminary version of this manuscript and assistance through all the stages of authors' work.

References

[1] Emelianova, N. A. (2006) Contribution of the professor Bellyarminov L. G. into glaucoma problem researching. *Glaucoma* 3, 77–80.

[2] Ivanidze, V. N. (2008). The Device for Light Exposure on the Human Body. Patent of Russian Federation No. 2359644.

[3] Markov, V. N. (2008). Colour Therapeutic Device. Patent of Russian Federation No. 2330693.

[4] Ohremenko, R. O. (1989). Peculiarirties of visual fatigue developing during the execution of precision work. *J. Ophthal.* 5, 272–275.

[5] Vinogradov, M. I. (1958). *Labour Process Physiology.* Cambridge: Publishing house of Leninrag university.

[6] Netudyhatka, O. U. (1987). The role of the critical flicker fusion frequency in the assessment of seafarers labor intensity. *J. Ophthalmol.* 5, 300–303.

[7] Peshkov, V. D. (1985). Relationship of subjective assessments of gymnasts with functional indicators. *Theory Pract. Phys. Cult.* 10, 11–13.

[8] Morozova, L. V., and Novikova, Y. V. (2013). Peculiarities of reading text from paper and electronic media Bulletin of Northern (Arctic) Federal University, Series. *Nat. Sci.* 1, 81–88.

[9] Portnih, U. I., and Makarov, U. M. (1987). Dynamics of CFFF depending on the focus of the training load. *Theory Pract. Phys. Cult.* 1, 46–47.

[10] Rosenblatt, V. V. (1975). "Problems of fatigue," 2nd edn., Rev. and add., M.: Medicine, 240.

[11] Kingsley P. (1995). Ancient Philosophy, Mystery and Magic. Empedocles and Pythagorean Tradition. Oxford, UK: Oxford University Press.

[12] Lurie, S. Ya Democritus. L. (1970).

[13] Da Vinci, L. (2007). A Treatise on Painting. Read Books.

[14] Newton, I. (1998). Opticks: or, a treatise of the reflexions, refractions, inflexions and colours of light. Palo Alto, Calif.: Octavo.

[15] Lomonosov, M. V. (1986). *Selected Works in 2 Volumes.* Moscow: Nauka.

[16] Young, T. (1802). Bakerian lecture: on the theory of light and colours. *Philos. Trans. R. Soc. Lond.* 92, 12–48.

[17] Helmholts, H. (1892). On the theory of compound colors. Phill. Mag. Vol. 4, 519–534.

[18] Brown, P. K. and Wald, G. (1964). Visual pigments in single rods and cones of the human retina. *Science,* 144, 45–52.

[19] Marks, W. B., Dobelle, W. H., and MacNichol, E. F. (1964). Visual pigments of single primate cones. *Science* 143, 1181–1183.

[20] Rushton, W. A. H. (1962). Visual pigments in man. Scientific American, 207, 120–132.

[21] Louise A. (1983). *Eye and Color.* Moscow: Energoatomisdat, 144.

[22] Hering, E. (1874). *Outlines of a Theory of the Light Sense.* Trans. L. M. Hurvich and D. Jameson. (Cambridge, MA: Harvard University Press).

[23] Hurvich, L. M., Jameson, D. (1956). Some quantitative aspects of an opponent-colors theory. IV. A psychological color specification system. *J. Opt. Soc. Am.* 46:6.

[24] Konig, A. (1987). *Ueber Blaublindheit.* Leipzig: Barth.

[25] Konig, A. (1903). *Ueber den Menschlichen Sehpurpur, und Seine Bedeutung fur Das Sehen.* Berlin: Sitzber, Akad. Wiss.

[26] Donner, K. (1957). "The spectral sensitivity of vertebrate retinal elements, Visual Problems of Colour," in *Proceedings of the National Physics Laboratory Symposium* (London: Her Majesty's Stationery Office).

[27] Hartridge, H. (1950). Recent Advances in the Physiology of Vision.—III. British Medical Journal 1(4666), 1331–1340.

[28] Bongard, M. M. Smirnov, M. S. (1957). Curves of spectral sensitivity of the receiver connected to singular fibrae of fog's optic nerve. Biophysics Vol. 2, 336–342.

[29] Vos, J. J., and Walraven, P. L. (1972). An analytical description of the line element in the zone-fluctuation model of colour vision. II. The derivation of the line element. *Vision Res.* 12, 1345–1365.

[30] Hubel, D. H. (1988). Eye, brain, and vision. New York: Scientific American Library.

[31] Land, E. H., McCann, J. J. (1971). Lightness and retinex theory. *J. Opt. Soc. Am.* 1971:61.

[32] Remenko, S. D. (1982). *Color and Vision.* Kishinev: Kartea Moldovenyaske 159.

[33] Marc, R. E. (2009). *Functional Neuroanatomy of Retina.* Available at: http://prometheus.med.utah.edu/~marclab/Marc_Duanes_ FNAR_2008 0815_layout.pdf//

[34] Schiffman, H. R. (2001). *Sensation and Perception: An Integrated Approach*, 55th edn. Hoboken, NJ: John Wiley & Sons.

[35] Satou, T., Ishikawa, H., Asakawa, K., Goseki, T., Niida, T., and Shimizu, K., (2016). Evaluation of relative afferent pupillary defect using RAPDx device in patients with optic nerve disease. *Neuro Ophthalmol.* 40, 120–124.

[36] Friling G., and Auer K., (1973) Man-colour-space.

[37] Mukhina V. S. (1981). *Child Art Activity as a form of Assimilation of Social Experience.* M.: Pedagogy.

[38] Yanshin, P. V. (2000). Color as a factor of mental regulation. *Appl. Psychol.* 4. 14–27.

[39] Kravkov, S. V. (1951). *Color Vision.* Moscow: Publish house AN USSR.

[40] Kitaev-Smyk, L. A. (1980). *Psychology of Stress.* Moscow: Nauka.

[41] Burlachuk, L. F. (1979). The Study of personality in clinical psychology. Kiev: Vishya Shkola, 176.

[42] Demidov, V. (1987). As we see what we see. Moscow: Znanie.

[43] Izmailov, C. A. (1995). Colour emotion characteristic. *Vest. MSU. Ser. psychol.* 4, 27–35.

[44] Goethe, I. V. (1957). Treatise about color // Selected works on natural history. M.

[45] Ovcharov, A. A. (2003). Children color diagnosis to determine the types of personality. *Psychol. Socionics Interpers. Relat.* 4, 31–38.

[46] Howard, N. Garb, James M. Wood, Scott O. Lilienfeld, M. Teresa Nezworski (Jan 2005). Roots of the Rorschach controversy. Clinical Psychology Review 25 (1), 97–118.

[47] Lüscher, M. (1971). *The Lüscher Color Test*, translation and ed. by Ian A. Scott. New York, NY: Pocket Books, 187.

[48] Bazima, B. A. (2001). *Color and the Mind (monograph)*. Kharkov, 172.

[49] Kravkov, S. V. (1950). Eye and its work. Moskow: USSR academy of sciences.

[50] Golubtsov, K. V., Kuman, I. G., Kheylo, T. S., Shigina, N. A., Trunov, V. G., Aydu, E. A.-I., et al. (2003). Flickering light in the diagnosis and treatment of pathological processes of the human visual system. *Inform. Process.* 3, 114–122.

[51] Christopher, W. Tyler and Russell, D. Hamer (1993). Eccentricity and the Ferry–Porter law. Journal of the Optical Society of America A Vol. 10, Issue 9, 2084–2087.

[52] Hecht, S., and Shlaer, S. (1936). Intermittent stimulation by light. V. The relation between intensity and critical frequency for different parts of the spectrum. *J. Gen. Physiol.* 19, 965–977.

[53] Krasnoperova, N. A. (1998). Critical flicker fusion frequency as an indicator of the development of fatigue of the deaf and visually impaired 6–9 years old children during a training exercise. *Defectology* 18–21.

[54] Egorova, T. S. (2002). Critical flicker fusion frequency in determining the visual performance of visually impaired students. Information processes/Egorova T. S., Golubtsov K. V. // (an electronic journal, www.jip.ru). Vol. 2, 106–110.

[55] Rozhentsov, V. V. (1996). Statistical evaluation of human functional state by the method of CFFF. Digital processing of multidimensional signals: materials of Russia scientific Conference, Yoshkar-Ola, 118–120.

[56] Kravkov, S. V. (1951). *Colour Vision*. Moscow: Publishing house of Science Academy USSR, 175.

[57] Rabkin, E. B. (2005). Polychromatic tables for research of color perception // Rabkin E. B. 11th edn. Mn: J. M. Sapozhkov, 56.

[58] Teterina, C. T., (1997). *Teterina's Device for Correction of Functional Systems of the Human Body*, Patent No. 2098059.

[59] Psyadlo, E. M. (1996). Physiological and hygienic assessment of pilots performance in the dynamics of the daily watch. *Med. Work Ind. Ecol.* 2, 34–37.

[60] Semyonovskaya, I. N. (1963). Electrophysiological studies in ophthalmology. Moscow: Medgiz, 279.

[61] Angeli, O., Veres, D., Nagy, Z., and Schneider, M. (2016). Reproducibility of measurements using the IMEA ADR III critical flicker-fusion frequency measuring device. *Orvosi Hetilap* 157: 1079.

[62] National Instruments, (2015). Official site, Available at: http://www.NI. com/ [accessed on 06 March, 2015.

[63] Tyagunin, A. V. (2015). About gender pecularities of dynamic vision of undergraduate students. *Tyagunin A. V., Morozova L. V. International student scientific bulletin 2.* Moskow: Akademiya Estestvoznaniya, 314–316.

4

EIGER Indoor UWB-Positioning System

**Jerzy Kołakowski[1], Angelo Consoli[2], Vitomir Djaja-Josko[1],
Jaouhar Ayadi[2], Lorenzo Moriggia[3] and Francesco Piazza[3]**

[1]Institute of Radioelectronics, Nowowiejska 15/19, 00-665 Warsaw, Poland
[2]Eclexys SAGL, Via dell'Inglese 6, 6826 Riva San Vitale, Switzerland
[3]Saphyrion SAGL, Strada Regina 16, 6934 Bioggio, Switzerland

Abstract

The EIGER project proposes a global approach for the efficient joint use of
Global Navigation Satellite Systems signals and ultra wideband technology
(UWB) positioning in order to allow permanent and reliable outdoor/indoor
localisation. This chapter is focused on the UWB-positioning subsystem
designed within the project. The architecture of the system and design of the
main devices are described. Results of selected laboratory tests of developed
equipment are presented and discussed.

Keywords: Indoor localisation, TDOA, UWB positioning.

4.1 Introduction

Nowadays, the technological progress creates an opportunity to provide
localisation-based services (LBS) in large outdoor and indoor areas. The most
popular localisation systems are Global Navigation Satellite Systems (GNSS).
Unfortunately, their signal is not accessible indoors. Therefore, in order to
supply the user with constant localisation service hybrid solutions should be
implemented. In these solutions, GNSS systems are supported by additional
localisation systems which are capable of working indoors [1, 2]. There are
several possible ways to enhance GNSS systems and allow them to work in the
indoor environments. Commonly used solution, A-GNSS relies on receiving
additional data allowing by the receiver (RX) to maximise both sensitivity

and acquisition speed which may result in better performance in places where GNSS signal level is low.

Another approach relies on the coupling GNSS systems with INS (Inertial Navigation System) ones. Such approach allows to track the RX even when GNSS signal is lost. It is possible by employing accelerometers and gyroscopes which readings can be used to calculate RX orientation and velocity and thus its location. However, such systems tend to decrease in accuracy over time [1].

The EIGER project focuses on the design of a GNSS/UWB-based system for indoor/outdoor localisation. The general EIGER system architecture is shown in Figure 4.1. In order to provide outdoor and indoor localisation, two separate localisation subsystems are used: the GNSS one and the UWB one. The navigation device consists of two modules: positioning module and a standard device (smartphone or tablet). The positioning module is intended for gathering location-related data, pre-processing, and sending data to the smartphone for further processing (e.g., location calculation and visualisation of the obtained position on the plan).

The positioning module can be used on its own in applications requiring recording tracking data. In this mode, localisation data is stored in an internal memory and can be post-processed using application running on an external computer.

Impulse Radio (IR)-UWB seems to be especially relevant for positioning purposes. The transmission of narrow pulses allows to measure time with high resolution (positioning techniques implemented in UWB systems usually

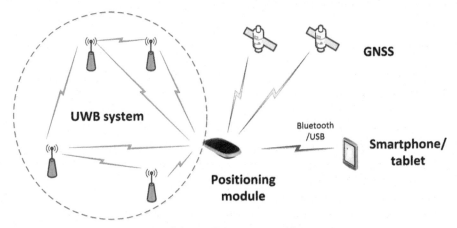

Figure 4.1   EIGER system architecture.

rely on measuring signal travel times between target and reference nodes). Additionally, the use of short pulses allows to increase systems immunity to interference and multipath propagation. Therefore, IR-UWB systems seem to be a good choice for indoor supplement of GNSS technology.

Hybrid GNSS-UWB solutions are the subject of several publications. The two main aims invoked by the authors are:

- increasing localisation accuracy and precision,
- providing localisation in both outdoor and indoor scenarios.

A technique for improving of GPS localisation accuracy of nodes in harsh forest environments by using additional information collected with UWB ranging system is presented in Hutchens et al. [3]. The authors propose a modified localisation algorithm taking into account distances between system nodes. The Levenberg–Marquardt algorithm was used for the square error minimisation.

Improvement of vehicle-to-infrastructure relative navigation accuracy is described in O'Keefe et al. [4]. The positioning method is based on results obtained from differential GPS system and UWB ranging solution.

Investigation of the UWB/GNSS localisation system in urban canyon environment is presented in MacGougan et al. [5]. The system consisting of two-way UWB radios with ranging capability and a GPS RX was tested in the area close to high buildings.

An interesting application of hybrid GPS/UWB positioning is proposed in Gross et al. [6, 7]. The Authors incorporate peer-to-peer UWB ranging results for UAVs (unmanned aerial vehicle) formation control during flight. Utilising information on distance between UAVs helped to improve positioning accuracy.

Combination of two technologies can also be used to locate vehicles in mixed indoor/outdoor scenarios (e.g., as automatic guided vehicles transporting goods in warehouses) with greater accuracy. In Fernandez-Madrigal and Gonzalez [8, 9], the particle filter localisation algorithm is proposed and investigated. A similar application was an objective of AGAVE project. The results can be found in AGAVE [10].

Described approaches can be classified into two main categories: tightly and loosely coupled. In tightly coupled techniques raw results obtained from both systems (e.g., pseudoranges, distances between nodes, information on satellite clock errors, or signal phase errors) are fused. The different versions of Kalman Filter are mainly the basis for those solutions. Unscented Kalman Filter applications are described in Gross et al. [6, 7], Extended

Kalman Filter-based solutions in Hutchens et al. [3], O'Keefe et al. [4], and MacGougan et al. [5]. An example of loosely coupled approach fusing results from UWB and RTK GPS (Real Time Kinematic GPS) systems is described in Khan et al. [11]. In all described solutions, UWB systems delivered information on ranges between nodes.

The UWB-positioning system, proposed in this chapter, is based on the TDOA (Time Difference of Arrival) technique. Selected TDOA implementation, in which the localised device receives and processes signals, has many advantages. The devices roles are strictly limited to the reception of GNSS or UWB signals. Therefore, the problem of intersystem interference is negligible. Moreover, such approach allows for limitless number of devices to be served simultaneously. In commercially available systems, where transmitting tags are localised, increasing number of tags can significantly decrease system performance.

TDOA systems include an infrastructure consisting of synchronised anchor nodes. Majority of out of the box systems use wire connections between devices in order to transmit synchronisation signals. In the presented system an innovative wireless synchronisation is introduced. The proposed transmission scheme utilises UWB signals for precise synchronisation and TDOA evaluation.

This chapter includes a description of the UWB part of the developed GPS/UWB system. The system architecture, nodes design, and the system operation is presented in Section 4.2. Next section includes a description of system investigation; it presents the results of static tests as well as tests consisting in moving object tracking. The discussion on positioning accuracy and precision is also included.

4.2 UWB-Positioning Subsystem

4.2.1 UWB System Architecture

The EIGER project addresses services intended for use in large areas, covering many distant buildings, often belonging to different institutions or enterprises. In such case, the UWB system infrastructure should be distributed. The UWB infrastructure consisting of many "islands" deployed in different premises is proposed. All these subsystems operate independently. Such approach significantly improves the system scalability. Extension of system operation area by "covering" another building becomes very simple. Another aspect encouraging system deployment is a low cost of infrastructure. Custom-designed transmission protocol allows for the same infrastructure

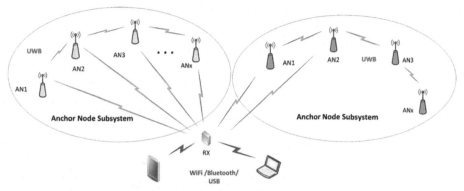

Figure 4.2 UWB system architecture.

to be shared between navigation devices offered by different operators. All these factors should enable wider system implementation. The system architecture, fulfilling above-mentioned requirements, is shown in Figure 4.2.

The infrastructures consist of anchors grouped into anchor node subsystems (ANSs). Although all of them share the same frequency range, they use different preamble codes in transmitted packets in order to limit inter-subsystem interference. Packets sent by the anchors included in the subsystem carry the same ANS Identifiers.

Each ANS is a set of wirelessly synchronised transmitters subsequently sending packets, each after a predefined delay. The RX receives packets and measures times of packets arrivals.

4.2.2 Anchor Nodes

The anchor node block diagram is presented in Figure 4.3. The UWB transmitter is based on Decawave's DW1000 integrated chip [16–18]. The UWB radio interface conforms to the IEEE 802.15.4a standard [12]. In order to provide chip clock frequency stability the external TCXO oscillator was used. The DW1000 chip cooperates with an UWB elliptical monopole antenna. The DW1000 transceivers are able to measure time of packet arrival with 15.65 ps resolution. They have also capability of triggering the transmission after the pre-programmed time.

The anchor node is controlled by Texas Instrument's TIVA family ARM Cortex M4 microcontroller. The node configuration is stored in internal microcontroller's EEPROM. The optional flash memory extends the capacity of this memory. It can be also used to store data transmitted by the anchors.

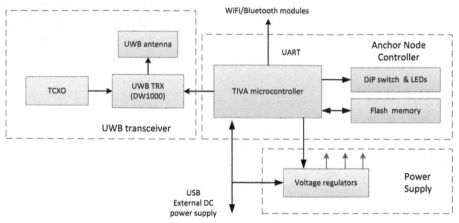

Figure 4.3   UWB module block diagram.

The anchor node is powered over the micro USB socket. The USB interface is used for setting node configuration.

UART interface makes the anchor node capable to be easily extended with external WiFi or Bluetooth transmission modules. By adding the transmission module and supplying the microcontroller with appropriate software it can be converted to the UWB test RX that allows investigation of the localisation system performance. The pictures of the anchor node board and the board with XBee WiFi module on top are shown in Figure 4.4. The test RX is a simplified version of the EIGER-positioning module including also GNSS RX. The software functionality controlling UWB system operation is the same in both designs.

The test RX measures times of packet arrivals and acquires data related to UWB signal levels. After collecting all UWB packets with the same sequence number (SQN) it sends all gathered data in one packet over the WiFi link to the PC.

The anchor nodes are supplied from the external source. Therefore, the current consumption is not of prime concern. However, in the battery operated RX high-energy consumption is a significant problem. To perform time of arrival measurements the DW1000 chip should be in the receive mode characterised by the highest supply current. Energy saving was achieved by switching the RX to idle mode in the time intervals between expected packets arrivals and by introducing sniffing mode. The mode is used after switching the device on or in case when the RX leaves the area covered by the system. In the sniffing mode the device periodically switches the RX on and checks for UWB signals availability.

(a)   (b)

Figure 4.4   Anchor node board (a) anchor node with WiFi module on top (b).

4.2.3 UWB Radio Interface

The EIGER UWB interface conforms to the IEEE 802.15.4a standard [12]. As the positioning system belongs to the Location and Tracking Type 1 (LT1) category, ETSI requirements concerning such systems [13] force its operation in 6–9 GHz frequency range. The DW1000 chip is able to work only in two channels in that range – 5th and 7th (the 5th channel was chosen in the system implementation). Both have the same centre frequency 6.4896 GHz but the signal bandwidths are different (500 and 900 MHz, respectively). Levels of UWB signals emitted by the transmitters fulfil requirements specified in regulations.

A general structure of the UWB frame is shown in Figure 4.5. It starts with a preamble after which a start of frame delimiter (SFD) is transmitted. According to the IEEE 802.15.4a specification [12] both parts form a synchronisation header. The system designer can choose between several preamble lengths ranging from 64 to 4096 symbols and a few preamble codes corresponding to different symbol sequences. The longest preamble transmission time can reach 4 ms.

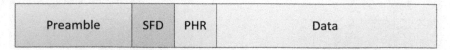

Figure 4.5   General structure of the UWB frame [12].

Table 4.1   UWB radio link parameters

Parameter	Value
Channel number	5 (6.4896 GHz)
Bandwidth	500 MHz
Preamble length	2048
Preamble codes	3 or 4
Packet data length	up to 127 bytes

The synchronisation header is followed by the frame physical header (PHR) including information on the frame format and its content. The data field carries localisation system information (e.g., identifiers and anchor coordinates) and may contain additional data related to positioning services (e.g., building plans). The detailed description of the frame content can be found in Kolakowski et al. [14].

The radio link parameters have a big impact on system performance (e.g., increasing the preamble length not only increases the transmission range, but also results in the increase in packet length and decrease in positioning rate). Therefore, a trade-off is needed. The parameters of UWB radio link implemented in the system are shown in Table 4.1. The time interval between transmitted packets is close to 10 ms. The transmission cycle repeats every 160 ms. However, those times can be easily tailored for custom needs.

4.2.4 Transmission Scheme

The EIGER UWB system is a TDOA-based solution. Therefore, the proper TDOA measurement is crucial to achieve good system accuracy and precision.

The developed transmission scheme had to achieve two goals:

- provide a way of anchor nodes synchronisation necessary in TDOA technique,
- enable TDOA measurements in the RX and deliver information supporting position evaluation.

4.2.4.1 Transmission scheme

Exemplary packet flow between anchor nodes (AN0 ... AN4) and the RX is shown in Figure 4.6. The packet transmitted by anchor n (marked as T_n) is received by all RXs and anchor nodes being within the transmission range (it is marked as R_n). In the presented example packet T0 is not received by anchor AN4.

The packet exchange starts with the packet sent by anchor AN0. All other anchors analyse received packet content, but only one responds with a delayed transmission of its own packet. All packets contain identifiers of transmitting anchors. The delayed transmission is triggered by the reception of the packet with the identifier of the preceding anchor in the transmission chain.

The malfunction of one anchor could break the transmission chain and block infrastructure operation. Therefore, the transmission scheme is equipped with a mechanism sustaining packet transmissions. Each anchor node stores identifiers of preceding anchors. After packet reception, the timer is started with the value depending on position of transmitting anchor in the chain. If another packet is received before the timer expires a new timer is started (this behaviour is illustrated by anchor AN2 operation). If no packets are received

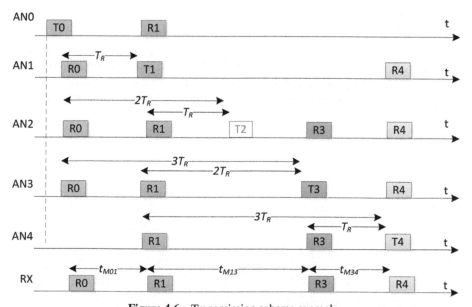

Figure 4.6 Transmission scheme example.

the transmission is triggered by the timer. Anchor AN3 behaviour illustrates this case.

Transmission from the last node in the ANS ends the transmission sequence. The whole described transmission chain is triggered periodically by anchor AN0.

4.2.4.2 TDOA calculation

The RXs measure time intervals between received packets and calculate TDOAs. The simplified transmission scheme illustrating the TDOA determination procedure is presented in Figure 4.7.

Packets sent by anchors reach the RX. It measures time intervals between consecutive packets, retrieves packet information and calculates TDOA values.

From Figure 4.7 the TDOA for AN0 and AN1 nodes is equal to:

$$\text{TDOA}_{\text{AN1AN0}} = t_{\text{PAN1RX}} - t_{\text{PAN0RX}}, \tag{4.1}$$

where t_{PAN0RX} and t_{PAN1RX} are propagation times of signal between anchors and the RX.

The time intervals defined in Figure 4.7 fulfil the following equation:

$$t_{\text{PAN0AN1}} + t_{\text{DELTXAN0}} + t_{\text{RXDELAN1}} + T_{\text{R}} + t_{\text{PAN1RX}} + t_{\text{DELTXAN1}}$$
$$+ t_{\text{DELRX}} = t_{\text{PAN0RX}} + t_{\text{DELTXAN0}} + t_{\text{DELRX}} + t_{\text{M01}}, \tag{4.2}$$

where: t_{PAN0AN1} is the propagation time between anchor nodes, T_{R}, predefined delay; t_{DELTXAN0}, delay in the AN0 transmitter; t_{DELTXAN1} and t_{DELRXAN1} are delays introduced by the transmitter and the RX

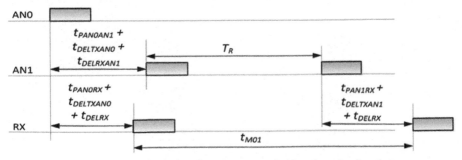

Figure 4.7   Time relations between transmitted and received packets.

in the AN1 node; t_{DELRX}, delay introduced in the RX, t_{M01}, measured time interval between packets from nodes AN0 and AN1.

After rewriting above equation the TDOA is equal to:

$$\text{TDOA}_{\text{AN1AN0}} = t_{\text{M01}} - T_{\text{R}} - t_{\text{PAN0AN1}} - (t_{\text{DELRXAN1}} + t_{\text{DELTXAN1}}).$$

(4.3)

The above equation specifies components required for TDOA determination. Besides the results of measurements performed by the RX, information on transmission delay (T_{R}) is necessary. The propagation time between the anchors can be either calculated from anchor nodes coordinates or measured. Measurement is justified in cases where inter-anchor communication takes place in NLOS conditions. Object obscuring propagation path can introduce delay or reflect the signal so the consecutive node is triggered by reflected (delayed) signals. The last component of the TDOA equation is a sum of delays introduced in the second node. Those values can be specific to particular anchors and should be determined by calibration measurements. The calibration procedure consists in measurement of TDOA in controlled environment, where distances between all nodes are known. The delay is obtained from Equation (4.3) after substitution of TDOA with the real value calculated from nodes coordinates. Performing a series of measurements and taking averaged TDOA value helps to reduce influence of random error components.

Several error sources contribute to the total uncertainty of TDOA derived from Equation (4.3). Measurement of time intervals and transmission delays are prone to anchor nodes' clock signal jitter, frequency tolerance, and instability. TCXOs used in the anchor design helped to reduce the problem. In order to check random TDOA measurement errors, the tests consisting in the reception of a series of packets transmitted by an anchor node were carried out. The RX measured time-of-arrival for incoming packets and calculated the TDOA values. Exemplary cumulative distribution function (CDF) of normalised results is shown in Figure 4.8. Ninety percent of results were measured with the error lower than 350 picoseconds.

4.2.5 Positioning Algorithm

The position determination algorithm is based on Extended Kalman Filter [15]. The filter calculates estimation of the state vector including RX coordinates and RX velocity components. In each iteration, the predicted state

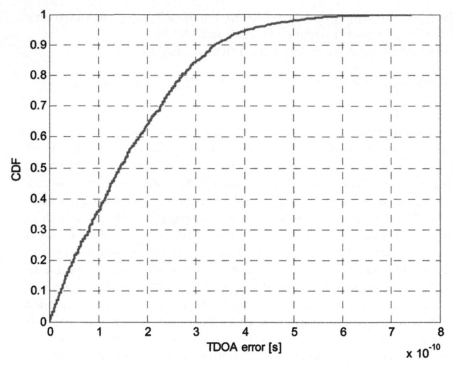

Figure 4.8   CDF of TDOA normalised measurement error.

vector values are corrected by taking into account the measurement vector-containing measured TDOA values. In the proposed filter implementation, the measurement vector length is not fixed, as it depends on the number of available TDOA results.

4.3 System Investigation

4.3.1 Test Scenarios

Tests took place in the hall of Department of Electronics and Information Technology building (second floor). Nine system anchors were mounted on the walls and tripods. The plan of the hall is shown in Figure 4.9. The hall is a demanding test environment because of several concrete pillars and some walls with metal cladding.

Two basic types of test scenarios were carried out. The first one depended on static measurements in several test points distributed over the hall area.

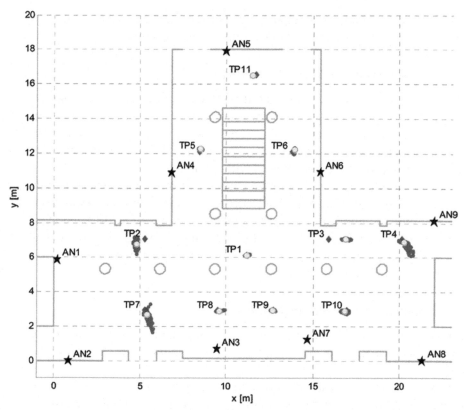

Figure 4.9 Calculated test points' positions.

The second one consisted in location of moving objects. In both experiments the test RX was attached to the small trolley at 1.3 m height. The trolley was parked in test points or pushed along predefined routes.

Tests were preceded by the system calibration performed in the P1 test point located in the middle of the area, in the place where LOS (Line-of-Sight) propagation to all anchor nodes was provided.

The collected results were analysed as follows. The accuracy and precision of localisation were determined for static measurements. The distance between the test point and the averaged position was a measure of the positioning error. The CEP (Circular Error Probability), corresponding to the radius of the circle in which assumed percentage of the results is included, was used for precision evaluation. The CEP was calculated for 50 and 90% of results.

In case of tracking tests the tracking error was characterised as a distance to the curve corresponding to the planned movement trajectory.

4.3.2 System Tests in Static Conditions

The locations of anchor nodes (marked with stars) and test points (marked with diamonds) are shown in the hall plan. At each point a few hundred measurements were performed. The positioning results are presented in Figure 4.9. Clouds of points correspond to locations calculated from particular results; averaged locations are marked with circles.

The best results are obtained in the central part of the area where signals from many anchors are transmitted in LOS propagation conditions. Away from the hall centre, the number of anchors' signals reaching the test RX is lower. Some of them are reflected from walls and pillars. Therefore, precision is worse and positioning error is larger. The summary of accuracy and precision measures is included in Table 4.2.

4.3.3 Localisation of Moving Objects

The results of moving object positioning are shown in Figure 4.10. The trolley with attached test RX was moved along two rectangular routes shown in the figure.

As in case of static tests the obtained tracking results are more accurate and precise in the hall centre and worse in side parts of the area. Multipath propagation is the main factor having impact on system performance. The signals from anchors were blocked not only by pillars and walls but also by the person pushing the trolley. The propagation conditions in the left part of the horizontal route were especially difficult, which resulted in higher positioning errors. These errors can be reduced by placing another anchor in this area. The tracking errors CDFs for both tracks are presented in Figure 4.11. In both cases 90% of track points were determined with an error lower than 0.5 m.

Table 4.2   Localisation accuracy and precision

Test Point Number	1	2	3	4	5	6	7	8	9	10	11
Error (m)	0.00	0.59	0.96	0.21	0.15	0.13	0.25	0.08	0.06	0.14	0.18
CEP 50 (m)	0.04	0.17	0.09	0.13	0.04	0.04	0.19	0.04	0.04	0.09	0.03
CEP 90 (m)	0.09	0.35	0.19	0.31	0.09	0.09	0.41	0.09	0.08	0.20	0.06

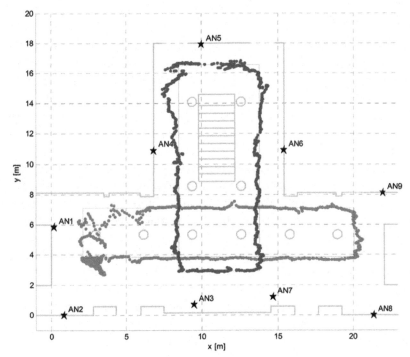

Figure 4.10 The RX tracking results.

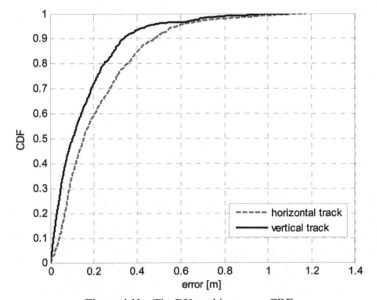

Figure 4.11 The RX tracking errors CDF.

4.4 Conclusions

Recent advances in technology resulting in availability of low-cost UWB transceiver chips create an opportunity to develop positioning systems competitive to commercial solutions. This chapter presents the indoor part of the UWB/GNSS localisation system developed within the EIGER project.

The system utilises a novel, wireless anchor node synchronisation technique. The solution lowers cost of the infrastructure and simplifies system deployment.

The system was tested in a demanding propagation environment. Investigation of the system performance in static conditions as well as tests consisting in moving objects tracking proved the quality of the proposed solution. Achieved positioning accuracy, better than 0.5 m, makes the solution sufficient to many applications.

The tests showed that multipath propagation is the main factor having impact on localisation accuracy. Careful planning of anchor nodes locations or increasing the number of nodes are the simplest solutions to these problems. In case of applications requiring better positioning accuracy and precision, data fusion of information obtained from the system and other sources (e.g., inertial MEMS sensors) can provide significant improvement.

Acknowledgements

The research leading to these results has received funding from the European Union's Seventh Framework Programme managed by REA – Research Executive Agency http://ec.europa.eu/-research/rea (FP7/2007_2013) under Grant Agreement No. 315435.

References

[1] Mautz, R. (2008). "Combination of Indoor and Outdoor Positioning," in *1st International Conference on Machine Control & Guidance*, ETH Zurich, 1–9.

[2] Reyero, L., and Delisle, G. (2008). A Pervasive indoor-outdoor positioning system. *J Netw.* 3, 70–83.

[3] Hutchens, C. L., Sarbin, B. R., Bowers, A. C., McKillican, J. D. G., Forrester, K. K., Buehrer, R. M. (2008). "An Improved Method for GPS-Based Network Position Location in Forests," in *IEEE Wireless Communications and Networking Conference* (Rome: IEEE), 273–277

[4] O'Keefe, K., Jiang, Y., Petovello, M. (2014). "An investigation of tightly-coupled UWB/low-cost GPS for vehicle-to-infrastructure relative positioning," in *2014 IEEE Radar Conference* (Rome: IEEE), 1295–1300. 10.1109/RADAR.2014.6875799

[5] MacGougan, G., O'Keefe, K., and Klukas, R. (2010). Tightly-coupled GPS/UWB integration. *J. Navigat.* 63.

[6] Gross, J., Gu, Y., and Dewberry, B. (2014). "Tightly-coupled GPS/UWB-ranging for relative navigation during formation flight," in *Proceedings of 27th International Technical Meeting ION Satellite Division*, Tampa Convention Center, Tampa, FL, 1698–1708.

[7] Gross, J. N., Gu, Y., Rhudy, M. B. (2015). Robust UAV Relative Navigation With DGPS, INS, and Peer-to-Peer Radio Ranging. *IEEE Trans. Automat. Sci. Eng.* 12, 935–944.

[8] Fernandez-Madrigal, J. A., Cruz-Martin, E., Gonzalez, J., Galindo, C., Blanco, J. L. (2007). "Application of UWB and GPS technologies for vehicle localization in combined indoor-outdoor environments," in *9th International Symposium on Signal Processing and Its Applications*, ISSPA, Nanjing, China, 1–4.

[9] Gonzalez, J., Blanco, J. L., Galindo, C., Ortiz-de-Galisteo, A., Fernandez-Madrigal, J. A., Moreno, F. A., Martinez, J. L. (2007). "Combination of UWB and GPS for indoor-outdoor vehicle localization," in *IEEE International Symposium on Intelligent Signal Processing*, WISP, Maryland, 1–6

[10] AGAVE. (2008). Project final report. Available at: http://cordis.europa.eu/documents/-documentlibrary/121407071EN6.pdf

[11] Khan, M. W. A., Lohan, E.-S., Piché, R. (2015). "Statistical sensor fusion of ultra wide band ranging and real time kinematic satellite navigation," in *2015 International Conference on Location and GNSS (ICL-GNSS)*, 1–6.

[12] IEEE Std 802.15.4a^{TM} (2007). *PART 15.4: wireless medium access control (MAC) and physical layer (PHY) specifications for low-rate wireless personal area networks (LR-WPANs) Amendment 1: Add Alternate PHYs.*

[13] ETSI EN 302 500 V2.1.1; ETSI EN 302 500-2 V2.1.1. (2010-10). *Location tracking equipment operating in the frequency range from 6 GHz to 9 GHz.* ETSI, Sophia Antipolis, France.

[14] Kolakowski, J., Consoli, A., Djaja-Josko, V., Ayadi, J., Morrigia L., and Piazza, F. (2015). "UWB localization in EIGER indoor/outdoor positioning system", in *Proceedings 2015 IEEE 8th International Conference*

Intelligent Data Acquisition and Advanced Computing Systems: Technology and Applications, Vol. 2, Warsaw, 845–849.

[15] Grewal, M. S., and Andrews, A. P. (2008). *Kalman Filtering: Theory and Practice Using MATLAB*, 3rd edn. Hoboken, NJ: Wiley.

[16] DecaWave. (2014). *DW1000 Data Sheet*, Dublin, Ireland: DecaWave.

[17] DecaWave. (2015). *EVK1000 User Manual, v.1.11*. Dublin, Ireland: Decawave.

[18] DecaWave. (2014). *DecaRanging (PC) User Guide, v.2.5*. Dublin, Ireland: DecaWave.

5

On Detection and Estimation of Breath Parameters Using Ultrawide Band Radar

Jan Jakub Szczyrek and Wiesław Winiecki

Institute of Radioelectronics and Multimedia Technologies,
Faculty of Electronics and Information Technology, Warsaw University
of Technology, ul. Nowowiejska 15/19, 00-665 Warsaw, Poland

Keywords: Ultrawideband radar, Data acquisition, Breath parameters, Real time systems, Noninvasive monitoring, Embedded systems.

5.1 Introduction

We develop and describe some algorithms dedicated to people characteristic movement-type detection using ultrawide band (UWB) radar. Derived methods are implemented in real-time system where breath detection and its basic parameters, e.g., frequency and depth are estimated. We also present here some crucial details of software implementation, where we deal with radar signal acquisition and processing in real time. Presented method has many interesting applications in healthcare: non-invasive respiration or heartbeat monitoring, and even through the wall measurement. Unlike camera, radar suspects anonymity. All development was made as Radcare project.

Movement recognition, breath detection, and also heartbeat estimation is well understood problem in literature [1, 2]. In many cases, narrow band microwave radars are used and in the most common situation Doppler effect is employed. But there are some problems mainly connected to strongly limited penetration of materials, especially in human body if fixed frequency is applied. There also exists a number of methods using cameras. But this way is uncomfortable and in many cases unacceptable by the patient.

In this chapter, we demonstrated new methodology using ultrawide band (UWB) Novelda NV6201 radar chipset built into embedded linux platform. The way algorithms are presented may be directly applicable in real life.

In our development, we do not assume and the position of the patient is known, so before measurement and estimation of bodily functions, we identify person position and then try to follow him using especially developed vital signs detection algorithm.

We apply our methodology on dedicated hardware and acquisition system dealing with samples in real time [3].

We provide two approaches. The first one is closely related to comparing radar impuls image with some periodically updated and filtered references, whereas the second method managing the signal shot by shot calculating cumulative differences. It needs to be underlined that typically some methods based on signal correlation is used. However, it is still applicable here, the cumultive differences consumes much less calculation power of processing unit. The final algorithm combines both algorithms using some references for breath rate estimation and calculating cumulative differences to get measured object position on the fly. Additionally, to fix estimated position, simple one-dimensional Kalman filter is applied.

First, results are obtained in Octave simulation software based on off-line calculation, while final algorithms are implemented on embedded system to produce on-line accessible results. This is no strict porting of the software from PC platform to embedded one because some additional preprocessing, compensation, and optimisation is needed in real-time system.

We also present our methodology to retrieve small signal of characteristic form from noise. Additionally, we show how to deal with observed thermal results and non-stationarity of measurements.

The radar system we are using has wide palette of parameters. We discuss some of the most important from our point of view. The deeper analysis of Novelda NV6201 radar functionality, construction, and parametrisation may be found in Piórek [4].

The software architecture of the system plays important role in our study and development hence we give a light of multi-thread architecture based on posix pthread Linux system library. The main portion of code was written in pure C language.

Presented algorithms are a part of Radcare system. The concept, modular structure, and building blocks of Radcare system is described in details in Sadkowski and Tajmajer [3], where one can also find what the system is dedicated for.

5.2 Data Acquisition and Preprocessing from UWB Radar

5.2.1 Signal Reprezentation

As a Novelda radar is of UWB type the signal has a form of pulse. The pulse received on antenna is next digitalised in a special way on AD converter, filtered and preprocessed in Novelda chipset (see [2]). Data samples are acquired from radar as a frames. The example of a sequence of 1000 consecutive frames is shown on Figure 5.1.

Theoretically – except some thermal fluctuation described later on – the situation is stationary and no movement takes place, i.e., there is nobody under observation. The graph is bolder at points where frames do not overlap exactly, which means that some noise or non-stationarity can be detected and in practise must be compensated.

The length of the frame served by Novelda chipset depends on parameter called *FrameStitch*. The minimum value is 256 samples which corresponds to 1 m in distance. The maximum is 15 m and we have choosen 2 m in our development. Additionally, an offset value may be specified. Which means that we have defined the start point and the distance in observed space.

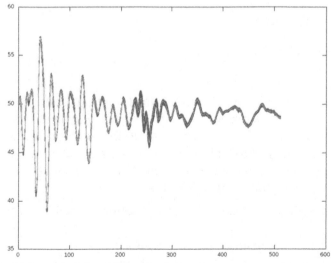

Figure 5.1 Radar signal for 1000 consecutive frames, start point 1 m and scanned distance 2 m.

The sequence of consecutive frames form a matrix M_{st} whose rows are defined by frames.

On the other hand we can look at frames as a function $f : S \mapsto \mathbb{R}$

$$f(s) = M_{st}|_{t=\text{Const}},$$

where domain S is a set of discrete points on a line

$$S = \{0, \ldots, m-1\}.$$

Similarly, if we define a set of discrete points of time

$$T = \{t_0, \ldots, t_{n-1}\},$$

then columns $r : T \mapsto \mathbb{R}$ are simply called traces $r_{s_*}(t)$

$$r_{s_*}(t) = M_{st}|_{s_*=\text{Const}}. \tag{5.1}$$

Typically, in our study, we need a bit more generalised definition of the trace. Namely, if in a place of fixed

$$s_* = \text{Const},$$

in Equation (5.1) we admit:

$$r(t) = M_{st}|_{s=s_0,\ldots,s_{m-1}}$$

with constrain:

$$|s_k - s_{k-1}| < C,$$

where $k = 1, \ldots, m - 1$ and C is predefined constant (natural number) close to 1. We call $r(t)$ generalised trace. One can easily imagine generalised trace like a path across matrix M_{st}, where we go from the top to the bottom jumping between columns but with limited deviation from left to right. This trace plays the crucial role in our development, hence it should be stabilised. To obtain this, we make following assumption:

- during the test the patient is fixed in nearly motionless, unknown position; and
- no other objects can disturbe scanned region.

According to this two assumption, we applied Kalman filtering to trace selection. This is done in one-dimensional situation. One can see in some aspects similar approach in two-dimensional situation in Wagner et al. [6].

5.2.2 Preprocessing

As it was mentioned before, some preprocessing may be done inside radar chipset and it depends on some internal parameters. The most interesting is parameter defined as *Iterations*. If iterations is equal to n the chipset serve each frame every n physical acquisition as a mean value:

$$F_t(s) = \sum_{\tau=0}^{n-1} f(\tau),$$

where t denotes the timestamp of completed results.

This not only acts similar to low-pass filter, but also divides frame rate. In our case, typically $n = 50$ which gives 100 frames/s. Observed that *Nyquist* frequency for traces, $r(t)$, is 50 Hz in this case. That gives enough room in frequency space because even extremal breath rate does not exceed 2 Hz. Performing some preprocessing outside central processing unit (CPU) allows to save calculation power for other, more sophisticated operations.

5.3 Off-Line Development

We started our research grabbing data directly from radar by use of PC software delivered by Novelda. This allowed us to understand some additional effects like slow thermal stabilisation which should be compensate. Off-line algorithm has two crucial steps:

- trace selection, which is roughly speaking, detection of the most significant movement along a line S,
- trace analysis, which is basically processing the function of this movement to determine vital signs.

5.3.1 Trace Selection

The first step in our development is to detect position of the patient. Formally for a given time sequence

$$T = \{t_0, \ldots, t_{n-1}\} \in \mathrm{T}$$

and given sequence of frames

$$F = \{F_{t_0}(s), \ldots, F_{t_{n-1}}(s)\}$$

we need to select s_* defining trace $r_{s_*}(t)$, which is optimal to estimate breath parameters. To obtain this, for each frame F_{t_i} we first compensate mean value

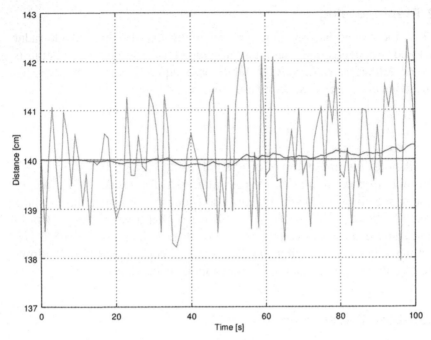

Figure 5.2   Simple Kalman filter applied to object located about 140 cm from radar antenna.

and select the most likely candidate for $r_{s_*}(t_i)$. Because of our assumtions, we expect that no urgent movement can occure. Hence, to reduce some thermal noise, we applied Kalman filter see Figure 5.2.

Next low-pass filtering is applied to form transformed acquisition matrix $\widetilde{M}_{st}$. Finally variation is calculated for each column s of this matrix, giving space variation estimate $v(s)$. Thus, we have

$$s_* = \max_{s \in S} v(s).$$

Figure 5.3 shows behaviour of space variation in case we have a person who is sitting at the front of the radar at distance approximately equal to 100 cm.

The person try to breath freely and stay in one position.

5.3.2 Trace Processing

Having s_* fixed, we are able to analyse the trace $r_{s_*}(t)$. First, we process it through low-pass filter with cut-off frequency 5 Hz. The result is shown on Figure 5.4. Selected trace gives a trajectory of radar signal at point s_* according

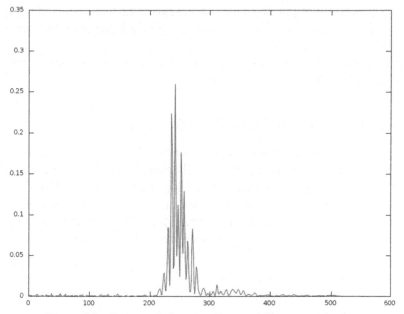

Figure 5.3   Space variation according to off-line trace selection.

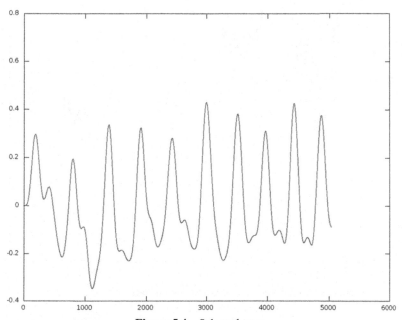

Figure 5.4   Selected trace.

to breath frequency and depth. Both parameters can be easily obtained directly from $r_{s_*}(t)$. We used Fast Fourier Transform to identify dominating frequency and its magnitude.

Observe that this trace has 5000 samples with 100 samples per second. Note that M_{st} has 5000 rows in this case.

5.4 Breath Detection in Real-Time System

Just after completion of off-line analysis, we decided to implement described algorithm on dedicated hardware platform [3] called RAD. The methodology cannot be moved directly; thus, we have developed some modifications.

During test sessions, we verified methodology and result. One of the most interesing observation was that respiration process can be detected even through the wall (Figure 5.5). That is another profit of UWB radar.

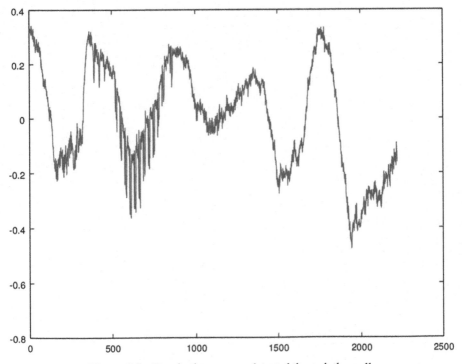

Figure 5.5   Respiration process detected through the wall.

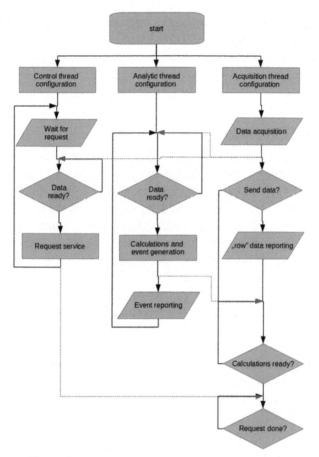

Figure 5.6 Block diagram of software airchitecture.

5.4.1 Sofware Architecture

RAD multicore platform is based on Linux operating system. Dedicated application must meet the following requirements:

- data acquisition must be done in real time;
- the application must perform necessary calculations in parallel; and
- the application is a source of data for other systems (e.g., visualisation).

All those reasons prompted us to develop multithread application which is able to make use of multicore hardware architecture.

Figure 5.6 shows the block diagram of the software. The application must run many tasks in paralell namely:

- Deriving bodility parameters and characteristics from collected data.
- Responding to some higher level command.
- Acquisition process parametrisation.
- "Raw" data reporting.
- Event reporting.

5.4.2 Movement Positioning

Processing M_{st} in real time system has some additional constrains. First of all, we need to identify position of the patient possibly quickly, thus less data is typicaly accessible to make decision. Especially M_{st} has much less rows then in theoretical case. On the other hand, taking s_* fixed we need to go back to historical data to reproduce trace. Finally, the system should follow patient position. Even if we assume that observed person is sitting, it is impossible to exclude any accidental body movement. Hence, typically traces should be replaced by generalised traces or a set of traces.

To calculate person position $s_*(t)$ as a function of time, we have developed special algorithm based on cumulated differences. This is done as follows: at the moment t first row of M_{st} is simply the newest frame, while the last is the oldest. Next, time rows are pulled down, the last row is removed and next newest acquisition is placed on position one. Thus, M_{st} has fixed size $m \times n$. Note that in this case typically $n = 16, \ldots, 64$ which is significantly less than 5000 which was a case in off-line method.

For each column $C_s(t)$ of M_{st}, we calculate

$$D_t(s) = \sum_{\tau=0}^{m-1} |C_s(\tau) - C_s(\tau - 1)|. \tag{5.2}$$

Intuitively $D_t(s)$ is a function that indicates points on the line S where the movement occur and it should correspond to the Equation (5.2). In practice, some additional low-pass filtering should be applied to $D_t(s)$ to get filtered difference $\widetilde{D}_t(s)$. This is strict consequence of the nature of human vital signs. Physical distance between points for used radar is about 4 mm. So picks on $D_t(s)$ should be removed. Figure 5.7 shows $D_t(s)$ for $m = 16$ and $m = 64$ and filtered difference. It can be easily shown that $\widetilde{D}_t(s)$ has well-defined maximum.

Assuming the patient does not move rapidly using $\widetilde{D}_t(s)$, we can assign generalised trace as

$$r(t) = \sup_{s=0,..,n-1} D_t(s)$$

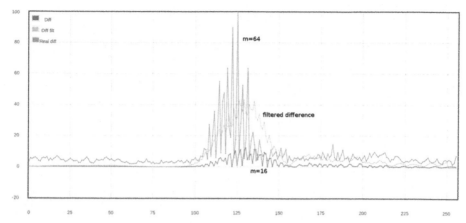

Figure 5.7 Human vital signs detection in real time. Filtered difference is the easiest to estimate maximum.

for each t. Next let us set

$$s_{\min} = \min_t r(t)$$

and

$$s_{\max} = \max_t r(t)$$

and two middle points

$$s_{\min} < s_0 < s_1 < s_{\max},$$

which allow us to draw Figure 5.8. Here, we have four traces $r_{s_{\min}}$, r_{s_0}, r_{s_1}, and $r_{s_{\max}}$ corresponding to four points spread on a line along distance 2 mm. All graphs are similar to sinusoid but have different phases. This effect is typical for respiration process as well as for other bodily functions (like heartbeat). Different phases corresponds to different body penetration by frequencies of UWB radar.

The frequency of the breath may be derived from FFT or using distance between cross points of graphs. To estimate the depth of the respiration, we use the magnitude of trace r_1 which is the closest one to generalised trace $r(t)$.

Let us underline that although positioning process plays crucial role in this development points s_0 and s_1 was choosen in some sense arbitrary and they behave nice. Thus, this means that the algorithm is not sensitive to exact positioning so much and is a very good information because practise patient may move slightly.

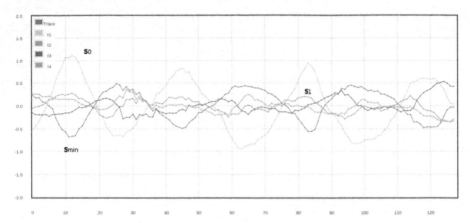

Figure 5.8   Four mutual close traces.

Another observation is closely related to the test described in Figure 5.4. Because respiration process may be detected and estimated though some obstacles (the wall in this example) the radar may be intergated under bed for vital signs detection.

All calculations are done in real time on embedded Linux system.

5.4.3 Suplementary Considerations

During our development some thermal instability radar signal was observed. This may affect some false movement detection especially if cumulative differences are taken over relative long time. Figure 5.9 presents radar signal instability during startup procedure when hardware is worming up.

Next picture (Figure 5.10) show this phenomenon in practise. Every 333 samples reference signal is taken from current acquisition:

$$R(s) = F_{t_0}(s).$$

Next the distance

$$\max_{s \in S} |F_{t_i}(s) - R(s)|$$

is calculated for $i = 0, \ldots, 333$.

That can be easyly seen that this thermal drift vanishes during the time. Hence, at the begining of measurement process some worming up procedure was implemented.

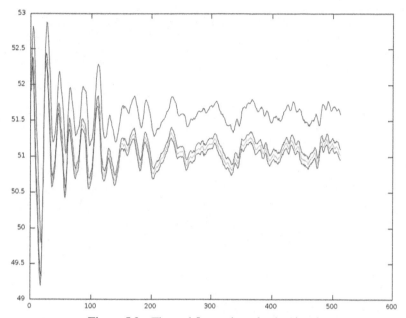

Figure 5.9   Thermal fluctuation of radar signal.

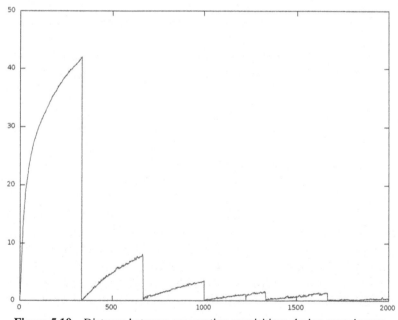

Figure 5.10   Distance between consecutive acquisitions during warming up.

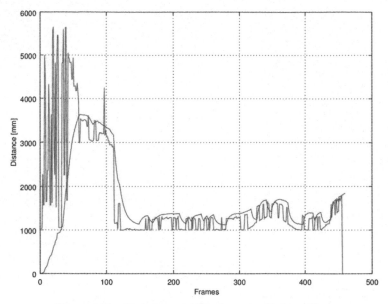

Figure 5.11   Trajectory stabilisation using Kalman filter.

As it was mentioned before Kalman filter was implemented for additional trajectory stabilisation. It is very useful if some instablility occurs as an effeect of reflections etc.

Figure 5.11 presents an example of stabilisation process in situation when patient change his position. Applied method allows to follow him with breath detection.

5.5 Summary

In this chapter, we present practical approach to respiration monitoring using UWB radar. Analytic background of our study as well as experimental results were presented.

Algorithms and methods we had developed were implemented on real-time platform based on embedded Linux system. This allowed us to perform on line patient position detection and respiration parameters estimation.

Described system defines first step in construction of non-invasive health-care type measurement platform. Construction of this platform is closely similar to construction of typical mobile systems, thus it seems to be potentially cost effective.

Acknowledgement

This work has been supported by EEA Grants: Norway Grants financing the project PL12-0001 (http://eeagrants.org/project-portal/project/PL12-0001).

References

[1] Fontana, R. J. (2004). Recent system applications of short-pulse ultra-wideband (UWB) technology. *IEEE Trans. Microwave Theory and Tech.* 52, 2087–2104.

[2] Lazaro, A., Girbau, D., and Villarino, R. (2010). Analysis of vital signs monitoring using an IR-UWB radar. *Prog. Electromag. Res.* 100, 265–284.

[3] Sadkowski, T., and Tajmajer, T. (2015). *CoAP and database integration for sleeping and non-routable nodes.* In: Proc. IEEE IDAACS Conference, Warsaw, Poland (available in Web of Science).

[4] Piórek, M. (2015). *On calibration and parametrization of low power ultrawideband radar for close range detection of human body and bodily functions.* In: Proc. IEEE IDAACS Conference, Warsaw, Poland (available in Web of Science).

[5] Bilich, C. G. (2006). "Bio-medical sensing using ultra wideband communications and radar technology: A feasibility study," in *IEEE Pervasive Health Conference and Workshops,* IEEE, Rome. doi: 10.1109/PCTHEALTH.2006.361671.

[6] Wagner, J., Miekina, A., Mazurek, P., Morawski, R. Z., Winiecki, W., Jacobsen, F., Øvsthus, K., Sudmann, T., Børsheim, I. (2016). "Signal processing in a two-module radar system for monitoring of human position and movements in an indoor environment," in *Proceedings of IEEE Conference on Signal Processing Algorithms, Architectures, Arrangements and Applications,* Poznañ, Poland.

6

Gabor-Filter-based Longitudinal Strain Estimation from Tagged MRI

Łukasz Błaszczyk[1,2], Konrad Werys[2,3], Agata Kubik[2,3] and Piotr Bogorodzki[2]

[1]Faculty of Mathematics and Information Science, Warsaw University of Technology, Warsaw, Poland
[2]Institute of Radioelectronics and Multimedia Technology, Warsaw University of Technology, Warsaw, Poland
[3]The Cardinal Stefan Wyszyński Institute of Cardiology, Warsaw, Poland

Abstract

Local contractility evaluation is a promising area of cardiovascular research. We propose and validate strain calculation method based on Gabor filters used on tagged magnetic resonance images. We focus on longitudinal strain calculated from images of the heart in its long axis. We validate the results on studies of both a healthy volunteer and two cardiomiopathy patients.

Keywords: Cardiac MRI, Motion tracking, Gabor filter, Motion quantification.

6.1 Theoretical Background

According to WHO [1], cardiovascular diseases are the first cause of death in developed countries and are projected to remain so. Evaluation of cardiac function is a crucial part of correct diagnosis and therapy planing. Visual assessment of the cardiac motion is possible using two non-invasive imaging modalities: echocardiography and magnetic resonance imaging (MRI). Evaluation usually involves qualitative part – visual observation by an experienced physician and quantitative part – calculation of global contractility

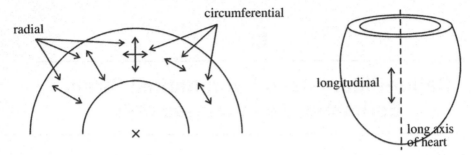

Figure 6.1   Schematic representation of the directions of the calculated heart strain.

parameters such as ejection fraction, and stroke volume. In many diseases there is a need of qualitative evaluation of local cardiac motion, but currently available methods are rarely used – either they provide unsatisfactory results or they need time-consuming post-processing. Due to high accessibility, echocardiographic methods such as Tissue Doppler or Speckle Tracking are more popular, though MRI provides better spatial resolution.

6.1.1 Tagged Magnetic Resonance Imaging (tMRI)

One of the typical perspectives in which a heart is observed in cardiac magnetic resonance (MR) is called *long axis*. In this perspective, cross-section of the heart through it's long axis can be observed (Figure 6.1). There are three typical long axis views: *two chambers*, where left ventricle and left atrium is visible, *three chambers*, where left ventricle, left atrium, and part of ascending aorta is visible (also known as left ventricular outflow tract view – LVOT), and *four chambers*, where both ventricles and both atria are visible [2]. The three views are presented in the figures in this chapter.

In tMRI, magnetic signal is locally attenuated, creating pattern of parallel lines [3]. The pattern moves with the tissue and enables its motion tracking. In this study, we use linearly increasing startup angles (LISAs) sequence [4] to get good temporal resolution (50 time frames), similar to speckle tracking echocardiography.

From mathematical point of view, tagging pattern may be described as a function $f_0(\mathbf{p})$ modulating tissue magnetisation at every material point with coordinates $\mathbf{p} = (p_x, p_y)^T$, which satisfies the condition $|f_0(\mathbf{p})| \leq 1$. Kerwin and Prince [5] showed that the sequence of N radio frequency (RF) pulses in classical spatial modulation of magnetisation (SPAMM) sequence is visible in k-space as $2N - 1$ peaks. The magnetisation modulation function (in time $t = 0$) has the form

$$f_0(\mathbf{p}) = \sum_{n=0}^{N-1} a_n \cos(n\mathbf{g}^T\mathbf{p}), \tag{6.1}$$

where vector $\mathbf{g}$ is the direction of the tagging gradient and coefficients a_n ($n = 0, \ldots, N - 1$) depend on the RF pulses' angles θ_ℓ ($\ell = 1, \ldots, N$) and can be approximated as

$$a_n \approx \begin{cases} 1 - \frac{1}{2} \sum\limits_{\ell=1}^{N} \theta_\ell^2 & \text{for } n = 0, \\ -\sum\limits_{\ell=1}^{N-n} \theta_\ell\theta_{\ell+n} & \text{for } n = 1, \ldots, N - 1. \end{cases} \tag{6.2}$$

Length of vector $\mathbf{g}$ determines the spatial frequency of the tagging pattern, while its angle determines tagging direction. Assuming that D and φ are required tagging distance and direction angle, respectively, vector $\mathbf{g}$ is given by the formula

$$\mathbf{g} = \frac{2\pi}{D} \begin{pmatrix} \cos\varphi \\ \sin\varphi \end{pmatrix}. \tag{6.3}$$

To simplify the notation, let us consider one-dimensional case with two RF pulses. Magnetisation modulation function is then formulated as

$$f_0(x) = a_0 + a_1 \cos(k_0 x), \tag{6.4}$$

where k_0 is the tagging frequency and coefficients a_0 and a_1 are determined by Equation (6.2). Denoting $I_0(x)$ as the value of the original image (without tagging), for the homogeneous heart tissue ($I_0(x) \equiv I_0$) we get

$$I(x) = I_0(x) \cdot f_0(x) = I_0(a_0 + a_1 \cos(k_0 x)). \tag{6.5}$$

From the definition of the complex cosine function ($\cos\varphi = \frac{1}{2}(e^{j\varphi} + e^{-j\varphi})$) we can also represent the above formula as

$$I(x) = I_0 \cdot \left(a_0 + \frac{a_1}{2}\left(e^{jk_0 x} + e^{-jk_0 x}\right)\right), \tag{6.6}$$

where j denotes the imaginary unit ($j^2 = -1$).

The Fourier transform of such signal is

$$I(k) = \frac{\pi}{2}I_0\left(a_0\delta(k) + \frac{a_1}{2}\delta(k - k_0) + \frac{a_1}{2}\delta(k + k_0)\right), \tag{6.7}$$

where $\delta(t)$ is Dirac delta function, which corresponds to three peaks in frequency domain. In general case, in k-space (which can be considered Fourier transform of the MR image) there are $2(N-1)$ peaks related to the $N-1$ cosine components and one peak related to the constant component [6]. In the real-life scenario, the heart tissue is not homogeneous, hence the peaks will be blurred.

During the tMRI examination tagging grid deforms in the same way the heart does. In order to describe the heart deformation we introduce the two-dimensional (2D) coordinate system related to the image in which every point is denoted as a vector $\mathbf{y} = (y_1, y_2)^T$ and the image value is described by $I(\mathbf{y}, t)$. Assume that $\mathbf{p}(\mathbf{y}, t)$ is a material point at coordinates $\mathbf{y}$ in the time moment t and $I_0(\mathbf{y}, t)$ is the image value at coordinates $\mathbf{y}$ and time t without any tagging. We need to remember that the tagging grid disappears with time, which can be represented by the factor $\beta(t)$ (a monotone function, decreasing from 1 to 0, dependent of the MR sequence). Finally, the sequence of tMRI images can by described by the formula:

$$I(\mathbf{y}, t) = I_0(\mathbf{y}, t) \cdot (\beta(t) f_0(\mathbf{p}(\mathbf{y}, t)) + (1 - \beta(t))). \qquad (6.8)$$

Such description is sufficient for the analysis of the tagging grid movement (and in effect for the analysis of all material points movement) in time. tMRI images are registered at regular intervals from the moment the tagging grid is imprinted. The resulting sequence of images is then processed.

6.1.2 Cardiac Strain

Due to geometry of the heart, cylindrical coordinate system (with z-axis aligned with axis of the heart) is best fitted for analysis of cardiac motion. Three components of the strain tensor are used: radial (E_{rr}), circumferential (E_{cc}), and longitudinal (E_{ll}) (as shown in Figure 6.1). Radial and circumferential components calculated from MRI were extensively researched [7–9]. Though proven useful in echocadriography [10], longitudinal component calculated from MRI did not get much attention.

In this chapter, we propose an automated approach for calculating left ventricular motion parameters. To assess strain information we use algorithm based on bank of the Gabor filters [11]. Frequency of a Gabor filter is adjusted so that is matches the local frequency of the tagging pattern. To validate our method, we analysed studies of three subjects: a healthy volunteer and two patients with cardiomiopathies.

6.2 Materials and Methods

6.2.1 MRI Sequence

MRI studies of a healthy volunteer and two patients with cardiomyopathy were acquired using 1.5 T scanner (Magnetom Avanto, Siemens). Tagged cine series of 50 images were acquired, using linearly increasing start-up angles (LISA) steady-state free precession (SSFP) SPAMM sequence [4], with retrospective pulse triggering. Imaging planes were set in the same way as in an echocardiographic study for the global strain measurement (two chambers, three chambers, and four chambers view) using following parameters: echo time $T_E = 1.3$ ms, effective repetition time $T_R = 34.1$ ms, flip angle FA = $20°$, slice spacing 6 mm, tag spacing 7 mm, and pixel size 1.07×1.07 mm.

For detection of a scar or fibrotic tissue in the patients, late gadolinium enhancement (LGE) sequence was acquired. In this method, first a contrast agent is injected. After some time ($\sim$10 min) the contrast agent is washed out from the healthy tissue but stays in the scarred or fibrotic tissue, causing signal enhancement as can be observed on Figure 6.2. LGE sequence parameters were as follows: echo time $T_E = 1.2$ ms, effective repetition time $T_R = 767$ ms, flip angle FA = $45°$, slice spacing 8 mm, and pixel size 1.48×1.48 mm.

6.2.2 Patient Data

A healthy volunteer and two patients with cardiomyopathy were studied. One patients suffers from dilated cardiomyopathy (DCM), a condition where the heart is enlarged, usually with myocardial wall thinning. The other patients suffers from hypertrophic cardiomyopathy (HCM), a condition where the myocardial wall are thickened and fibrotic.

6.2.3 Longitudinal Strain Estimation Using Gabor Filter Bank

Our method utilises the fact that in tMRI technique the tagging lines are imprinted on the myocardium and can be set to be perpendicular to the heart's long axis, thus perpendicular to the longitudinal strain direction. The tagging lines, parallel and equally distant, are 'imprinted' at the end diastole and deformed according to the heart motion. As a result, relative difference between the distance of the lines in each material point at the beginning (ℓ_0) and at the end of deformation (ℓ_1) is equal to the longitudinal strain $\varepsilon_{\ell\ell}$, as in the formula:

$$\varepsilon_{\ell\ell} = \frac{\ell_1 - \ell_0}{\ell_0}. \tag{6.9}$$

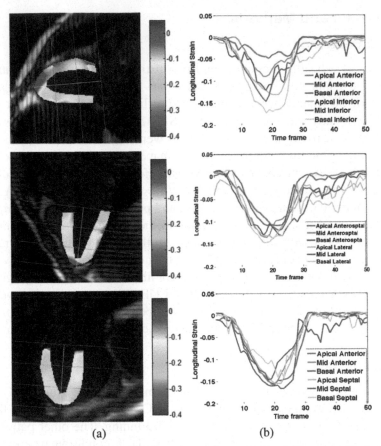

Figure 6.2 Long axis view of a healthy volunteer heart. (a) Tagged images with strain maps in the end systolic phase. (b) Curves of mean longitudinal strain per cardiac segment.

To calculate a local distance between tagging lines we use a bank of Gabor filters [11]. This method was successfully used by Qian et al. [12] to calculate radial and circumferential strain in the short axis tagged images.

A Gabor filter is a band-pass filter, its impulse response is given by the formula

$$g(x, y) = s(x, y) \cdot w_r(x, y), \tag{6.10}$$

where w_r is a 2D Gaussian function (called envelope) and s is a complex 2D sinusoid (carrier). Impulse response carrier is defined as

$$s(x, y) = \exp(j(2\pi(u_0 x + v_0 y) + \varphi)), \tag{6.11}$$

where (u_0, v_0) is 2D spatial frequency and φ is its initial phase. Gaussian envelope is described as

$$w_r(x, y) = K \cdot \exp(-\pi(a^2(x - x_0)_r^2 + b^2(y - y_0)_r^2)), \qquad (6.12)$$

where (x_0, y_0) is spatial translation of a Gaussian function, a and b define its width and K is a scaling factor. The lower indice r defines the rotation of a Gauss envelope around the center of the coordinate system at an angle θ:

$$(x - x_0)_r = (x - x_0) \cos \theta + (y - y_0) \sin \theta, \qquad (6.13)$$

$$(y - y_0)_r = -(x - x_0) \sin \theta + (y - y_0) \cos \theta. \qquad (6.14)$$

The scaling factor K is usually assumed to be $a \cdot b$. Parameters x_0, y_0, and φ are also usually set to 0.

Frequency response of a Gabor filter is then given by the formula:

$$G(u, v) = \exp\left(-\pi\left(\frac{(u - u_0)_r^2}{a^2} + \frac{(v - v_0)_r^2}{b^2}\right)\right). \qquad (6.15)$$

It is worth noticing that the frequency response of the Gabor filter is a Gaussian function shifted by the vector (u_0, v_0).

Consider the fact that the tagged image is a traditional MR heart image modulated by a sinusoidal tagging pattern. We use a bank of the Gabor filters with different frequency parameters. It turns out that the Gabor filter with frequency parameter (u_0, v_0) equal to the local spatial frequency of tagging at point **p** gives the highest response (with respect to the amplitude) at this point.

In our algorithm, we directly use the bank of Gabor filters (with set of parameters predefined to cover most of the possible directions and distances of tagging) and for each pixel of the image we pick the filter that maximises the response, and get its main frequency vector (u, v). By simply taking inverse of the norm of such vector, as in formula:

$$\ell_1 = \frac{1}{\sqrt{u^2 + v^2}}, \qquad (6.16)$$

we get local tagging distance. By projecting this distance to the long axis of the heart and calculating the relative change of this distance, we get the value of longitudinal strain. Estimated calculation time was about 30 s per one image series.

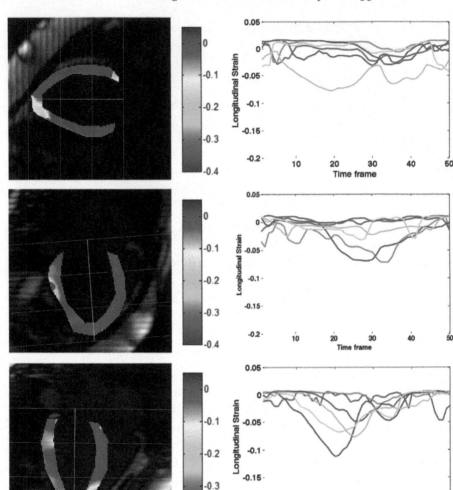

(a)   (b)

Figure 6.3   Long axis view of a DCM heart. (a) Tagged images with strain maps in the end systolic phase. (b) Curves of mean longitudinal strain per cardiac segment.

6.3 Results

The myocardium was segmented and cardiac regions were selected according to the American Heart Association guidelines [2]. Longitudinal strain was calculated for each pixel, strain curves were calculated as a mean value of strain

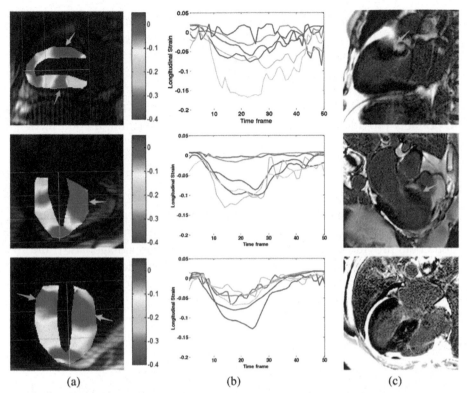

(a)   (b)   (c)

Figure 6.4 Long axis view of a hypertrophic cardiomiopathy heart. (a) Tagged images with strain maps in the end systolic phase. (b) Curves of mean longitudinal strain per cardiac segment. (c) LGE images. *Green arrows* indicate low strain regions corresponding to LGE enhancement. Blue arrows show low contractility in regions with no LGE enhancement.

from all pixels in a cardiac region. Results from a healthy volunteer study are presented in Figure 6.2. Similarly, results from patients with cardiomyopathies are presented in Figures 6.3 and 6.4. In one patient fibrotic tissue was detected on LGE images, as presented in Figure 6.3.

6.4 Discussion

Strain maps in a healthy volunteer study are smooth and uniform, with high strain values. Strain curves look as expected with clearly visible peak at the end of systole. In the first patient, suffering from DCM, smaller values of strain maps show much smaller values in the whole myocardium region. Values on the curves show significantly smaller strain than in the volunteer, with no visible end-systole peak. No fibrotic tissue could be observed in LGE images (not included). In the second patient, suffering from HCM, we observe regions with lower strain values. Some of these regions, indicated by green arrow, correspond to fibrotic regions in LGE images.

Strain maps and curves were consulted with physicians. The results from our method agree with the findings in disease history of the patients and could possibly provide additional insight in disease evaluation, if confirmed on bigger group of patients.

6.5 Conclusion

A novel, fast method to calculate longitudinal strain was proposed. Further work is needed to confirm our findings and extend the method to 3D. The method could provide additional insight to cardiac diseases diagnosis. If the findings are confirmed, this method could provide information about scarred or fibrotic tissue without use of contrast agent, making cardiac MRI study both easier for patients and more economic for the hospital.

References

[1] WHO. (2014). *The Top 10 Causes of Death*. Geneva: WHO.

[2] M. D. Cerqueira. (2002). Standardized myocardial segmentation and nomenclature for tomographic imaging of the heart: a statement for healthcare professionals from the cardiac imaging committee of the council on clinical cardiology of the American Heart Association. *Circulation* 105, 539–542.

[3] Zerhouni, E. A., Parish, D. M., Rogers, W. J., Yang, A., and Shapiro, E. P. (1998). Human heart: tagging with MR imaging–a method for noninvasive assessment of myocardial motion. *Radiology* 169, 59–63.

[4] Zwanenburg, J. J. M., Kuijer, J. Tim Marcus, J., and Heethaar, R. M. (2003). Steady-state free precession with myocardial tagging: CSPAMM in a single breathhold. *Mag. Reson. Med.* 49, 722–730.

[5] Kerwin, W. S. and Prince, J. L. (2000). A k-space analysis of MR tagging. *J. Magn. Reson.* 142, 313–322.

[6] Kerwin, W. S., Osman, N. F., and Prince, J. L. (2008). "Image processing and analysis in tagged cardiac MRI," in *Handbook of Medical Image Processing and Analysis*, I. N. Bankman ed., 2 edn., Chap. 26, 435–452. Cambridge, MA: Academic Press.

[7] Petryka, J., Miśko, J., Przybylski, A., Śpiewak, M., Małek, L. A., Werys, K., Mazurkiewicz, L., Gepner, K., Croisille, P., Demkow, M., and Rużyłło, W. (2012). Magnetic resonance imaging assessment of intraventricular dyssynchrony and delayed enhancement as predictors of response to cardiac resynchronisation therapy in patients with heart failure of ischaemic and non-ischaemic etiologies. *Eur. J. Radiol.* 81, 2639–2647.

[8] Bilchick, K. C., Kuruvilla, S., Hamirani, Y. S., Ramachandran, R., Clarke, S. A., Parker, K. M., Stukenborg, G. J., Mason P., Ferguson, J. D. Moorman, J. R., Malhotra, R., Mangrum, J. M., Darby, A. E., Dimarco, J., Holmes, J. W., Salerno, M., Kramer, C. M., and Epstein F. H. (2014). Impact of mechanical activation, scar, and electrical timing on cardiac resynchronisation therapy response and clinical outcomes. *J. Am. Coll. Cardiol.* 63, 1657–1666.

[9] Sohal, M., Duckett, S. G., Zhuang, X., Shi, W., Ginks, M., Shetty, A., Sammut, E., Kozerke, S., Niederer, S., Smith, N., et al. (2014). A prospective evaluation of cardiovascular magnetic resonance measures of dyssynchrony in the prediction of response to cardiac resynchronization therapy. *J. Cardiovasc. Magn. Reson.* 16, 58.

[10] Reisner, S. A., Lysyansky, P., Agmon, Y., Mutlak, D., Lessick, J., and Friedman, Z. (2004). Global longitudinal strain: a novel index of left ventricular systolic function. *J. Am. Soc. Echocardiogr.* 17, 630–633.

[11] Movellan, J. R. (2002). *Tutorial on Gabor filters. Open Source Document.*

[12] Qian, Z., Liu, Q., Metaxas, D. N., and Axel L. (2011). Identifying regional cardiac abnormalities from myocardial strains using nontracking-based strain estimation and spatio-temporal tensor analysis. *IEEE Trans. Med. Imag.* 30, 2017–2029.

7

A Decision Support System for Localisation and Inventory Management in Healthcare

**Francesca Guerriero[1], Giovanna Miglionico[2]
and Filomena Olivito[1]**

[1]Dipartimento di Ingegneria Meccanica Energetica e Gestionale,
Università della Calabria, Italy
[2]Dipartimento di Ingegneria Informatica, Modellistica,
Elettronica e Sistemistica, Università della Calabria, Italy

Abstract

In the last years, the Italian healthcare system is facing significant challenges, which underline the serious need for radical changes. Indeed, there is broad evidence that, even though a huge amount of money is invested in healthcare services every year, the patients often do not get the care they need. Thus, reforming the healthcare delivery system is mandatory, to ensure high-care quality at reasonable cost. It is evident that Information Technologies can play a central role in supporting decision makers in improving healthcare services. In this work, we present a Decision Support System (DSS) embedding localization and inventory optimisation models defined to address two types of problems arising in healthcare. The first is of strategic type and refers to hospitals' localization and healthcare service network reorganization. The second is of operational type and it deals with the inventory management of hospital pharmacies and departments, that are part of the current network. The proposed DSS can be an useful tool to take decisions both at strategic and operational level.

Keywords: Decision Support System, Localisation Problem, Inventory Management Problem, Calabrian Healthcare Network, Reorganization Models.

7.1 Introduction

This chapter presents a Decision Support System (DSS) developed to support decision makers, dealing with management problems in healthcare. The management problems are related to both strategical and operational issues and more in details to facilities location and inventory management problems. The growing attention to the measuring of performance and the necessity of a rationalisation of the healthcare resources [1–5] make mandatory the use of DSS in healthcare [6–8]. Indeed, several products are available to support healthcare decision makers at different levels like INSIGHTS [9] developed for financial and operational management in healthcare institutions; the Qlik platform that enables healthcare institution to explore clinical, financial, and operational data to find insights which lead to greater efficiency [10] and [11] which offers several solutions for inventory management issues.

It is well-known that the choice of the place, where facilities should be located, strongly affects the quality of the delivered service. This issue is particularly important in the public sectors: indeed, the localization of hospitals, schools, and emergency services greatly influences the accessibility of customers and, consequently, the level of service provided.

Given the practical importance of these decisions, several models and methods have been published in the scientific literature to address different variants of the facility location problem [12–14].

The developed models can be grouped in the following three main categories: *median, covering*, and *centre* models.

The models belonging to the first group (i.e., median models) are aimed at determining the optimal location of an *a-priori* fixed number of facilities, in such a way that the total distance between customers and their assigned facilities is minimized [15, 16].

In the covering models, the so-called coverage radius is introduced to discriminate among the facilities. Indeed, a demand point can be assigned to a given facility only if their distance does not exceed the coverage radius.

Two different variants of the covering problem have been addressed. In the former (referred to as set covering location problem), the objective is to minimize the number of facilities required to satisfy the entire demand in a given area [17], while in the latter (referred to as maximum covering location problem), the aim is to maximize the satisfied demand, under a limited number of available facilities [18].

The main aim of the centre models, originally introduced by Hakimi [15], is to locate the facilities, in such a way that each demand point receives the service from the closest facility and the maximum distance between each demand point and its assigned facility is minimised.

It is important to underline that, sometimes, in real settings, some modifications of the configuration of an existing facility network are required, to guarantee an adequate level of service, when some critical parameters (i.e., demand, costs, market structure, and degree of competition) undergo significant changes. In this case, a reorganization of the network takes place, involving changes in the number or the position of facilities, their capacities and/or types of supplied services.

In the scientific literature, different reorganisation models have been developed by considering two different situations: the reorganisation takes place before the parameters variations occur (ex-ante), on the basis of forecasting the future [19, 20]; or alternately, the modifications of the network are implemented ex-post, that is after the changes have happened [21].

Inventory management in healthcare has the same basic motivations as those faced by all the other organisations with this problem: namely, the availability, whenever necessary of the resources that guarantee an appropriate level of assistance; the reduction of costs generated from high inventory and/or from expired drugs; the avoidance of drug shortages especially vitally important ones.

The practical importance of the considered problem is proved by the literature focus on both different aspects of the inventory management, including supply chain operational and general management issues [22–26], and on real case studies (for example [27, 28]).

General inventory management and control have attracted the attention of many researchers over the years and there is a huge body of related literature. Among others, the reader can refer to Ghiani et al. [29], Waters [30], and Johnson and Montgomery [31] for inventory management motivations together with the introduction of different optimisation models and the related solution approaches.

In this work, we focus on the definition of a DSS based on some optimisation models, to mathematically represent the problems arising in the localisation of public facilities (i.e., hospitals); in the reorganisation of the regional public healthcare network; and in the inventory management. The models embedded into the DSS are described and implemented with reference to a real case in two papers. In particular, for the localisation and reorganisation models the readers is referred to Guerriero et al. [32], and for a detailed description of the implemented inventory model and policies, the reader is referred to Guerriero et al. [33].

The rest of the chapter is organized as follows: Section 7.2 describes the methodologies we use in designing the DSS, while in Section 7.3 the models, that represent the optimization core of the DSS are introduced; in Sections 7.4 and 7.5 the functionality of the system are presented and discussed in details; we draw some conclusions in Section 7.6.

7.2 Methods

The architecture of the implemented DSS for healthcare management is presented in Figure 7.1.

The DSS can be used in two different modalities.

The first one is designed for "expert optimization" users and offers them the possibility of solving both the localisation and the inventory models described in Section 7.3.

This functionality requires a good knowledge of the optimization models embedded in the system and it will be discussed in details in Section 7.5, with reference to the inventory management problems.

The second modality is designed for "no-expert users" and gives them the possibility of making a "what-if analysis", by varying some of the parameters of the current hospitals network configuration. This modality is characterized by a graphical user-friendly interface, that allows the user to insert the initial data easily and to analyse the output optimisation process very quickly.

The applications are implemented in Java Server Faces (JSF), the integrated optimization models in Cplex 12.1, while the database management system is MySQL. The system can be used on the web and it runs on JBOSS 7.1.

Once an authenticated session has been established at the site (http://185.54.154.40:9000/tess/) trough the *Access Control*, the *Interface Module* becomes available giving the possibilities to users of using the DSS in both "expert" and "no-expert users" modalities. The core of the system is the *Optimisation Module* that, once the necessary data are introduced, solves the related mathematical models and provides the solutions that are the base of the decision making process.

A very useful component of the system is represented by the *Google Maps* API, that gives a consistent and precise visualisation of the hospitals' localisation in the region of interest. In addition, it offers a set of functionalities, that facilitate some of the operations the users can perform on the actual configuration.

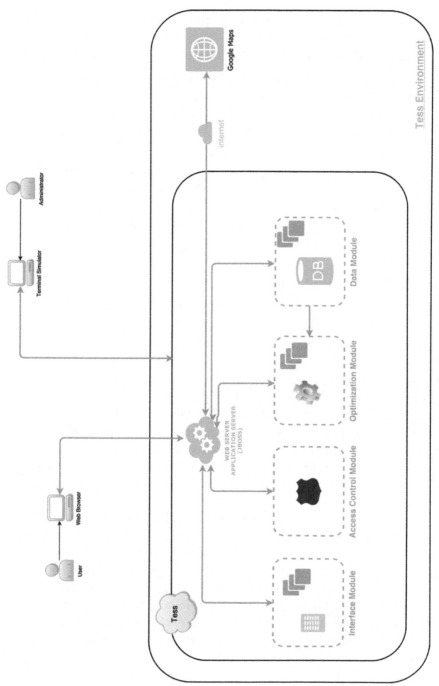

Figure 7.1 System architecture.

The communication with the database is performed by the *Data Access* system.

Summarising the exchange of data and information is made through:

1. the user interface to facilitate the input of the data and the analysis of the generated solutions;
2. the business logic constituted by the optimization solver;
3. the relational database for the storage and the retrieval of the data.

The expert users modality gives the possibility of using the optimisation models for localisation and reorganisation and inventory management. The "no-expert users", designed to help management in strategic issues, is related to localisation and reorganisation.

In the next section, we introduce the main features of the optimisation models used in the implemented DSS.

7.3 The DSS Optimisation Models

In this section, the optimization models, embedded in the proposed DSS, are shortly introduced. The detailed description, mathematical formulation, and implementation of the models used to solve both the localisation and the reorganisation problems in case of the Calabrian region can be found in Guerriero et al. [32], and the models and policies used for optimal inventory management are discussed in Guerriero et al. [33].

With reference to the localization problems, in the proposed DSS a classical p-median formulation is implemented and solved [15]. The formulation aims at locating p hospitals, while assigning the demand from the demand-centres to the departments, with the objective of minimising the total network travel time. Constraints on demand satisfaction and departments capacity are also imposed.

The reorganisation model embedded in the DSS is aimed at improving the efficiency and to rationalise the hospitals capacity (in terms of number of beds), in order to meet national care standards. As a matter of fact, the number of spokes to be activated should be minimised, while guaranteeing an appropriate level of service (measured in terms of satisfied demand) to the customers and considering the national and local authorities requirements. The main aim of the considered mathematical formulation is to select the hospitals to be activated, representing the spokes of the network, by taking into account the regional reorganisation guidelines.

In particular, the following specific aspects have been taken into account when formulating the problem of the healthcare network reorganisation:

- the total number of available beds, in the considered region, has been determined by considering a "target" number of 4 beds per 1000 inhabitants;
- for each department, the total number of beds available by considering all the chosen spokes should be greater than a given threshold: this guarantees that the territorial demand for specific services is satisfied;
- a given capacity (i.e., total number of beds) should be available in each spoke;
- a minimum number of beds should be assigned to each department;
- specific departments need to be present in all activated hospitals.

By setting some parameters, the mathematical models, briefly described above, can be used to first obtain an initial scenario for the healthcare network and successively to carry out a "what-if" analysis, by varying them through the DSS "no-expert users" functionalities, described in the Section 7.4.

With reference to the inventory management, the main aim of the optimisation models, implemented in the DSS, is to define the cost reduction strategies.

In particular, we implemented the Multi-product Capacitated Inventory Problem (MCIP) [32] aimed at minimising the inventory and ordering costs and its three variants designed to minimise the total number of orders on the planning horizon T.

In particular, the first variant is focused on the minimisation of the overall number of orders on a given planning horizon. The second variant encourages an order distribution, where the number of orders for each product is minimised over the planning horizon; finally the third variant minimises the total inventory level on the planning horizon.

The implemented models represent very important tools to support the decision makers, who can select the most suitable one, depending on the specific features to take into account. In addition, they can be used to analyse the behaviour of the inventory system under different operational conditions. In defining the above models, we have considered department and hospital pharmacies as separate parts of the same system: they decide autonomously the drug quantities to be ordered and when the order has to be placed. The suppliers for the hospital pharmacies are the drugs producer, while the consumers are the department pharmacies; on the other hand, the supplier for department pharmacies is the hospital pharmacy and the consumers the department patients.

We also defined a new mathematical model aimed at taking into account possible synergies between the two types of pharmacies. In particular, we introduced the Integrated Inventory Problem (IIP) [32] with the objective of minimising the total inventory and dispatching costs from the hospital to the department pharmacies. The design requirement is that the hospital pharmacy is a primary depot with the quantity of drugs necessary to satisfy all the demand from the department pharmacies.

To take into account stochastic and stationary demands, we also implemented in the DSS the well known Fixed Order Quantity and Fixed Period inventory policies. Under the *Fixed Order Quantity* (FOQ) policy, the order of a fixed quantity Q is made every time the inventory goes below a predetermined value s, called *reorder point*. Under the *Fixed Period* (FP) policy, an order is made with a fixed frequency. The quantity to be ordered is variable and such that, when the order is made, the inventory grows to the value of S (the *order-up-to-level*) so that the demand (until the arrival of the next order) is satisfied with probability α.

The introduced policies represent valid tools to support the decision makers in defining inventory management strategies, aimed at costs reduction. The proper use of the discussed inventory models is determined by the nature of the decisions to be taken and by the available information on the demand, but it is clear that an important increase in efficiency can be obtained by integrating them into a dedicated decision support system.

7.4 The "No-Expert Users" Functionalities of the DSS

We describe some of the functionalities of the DSS and in particular some of the issues relevant for location and reorganization problems in healthcare.

In this section, we discuss the modality designed for "no-expert users" that gives to decision makers the possibility to take decisions on the basis of a "what-if analysis" and by the comparison of different scenarios, obtained by varying some of the parameters that define the initial network configuration.

Once the user authentication is made and the "no-expert users" modality is selected, the user can access, from the home page, the following features:

- Localization (Programmazione in Figure 7.2): the decisions to be made are based on the localisation model introduced in Section 7.3 and discussed in detail in Guerriero et al. [32]. It is possible to compare a default scenario of the current hospital network (Initial Scenario) with a new scenario (Post-Elaboration Scenario) obtained by varying some of the network parameters like the number of hospitals, the number of the departments, and the healthcare service demands.

- Reorganization (Progettazione in Figure 7.2): the decisions to be made are based on the reorganisation model introduced in Section 7.3 and deeply discussed in Guerriero et al. [32]. In particular, it is possible to compare a default scenario of the hospital network (Initial Scenario) with a new scenario (Post-Elaboration Scenario) obtained by varying some of the parameters like the minimum number of beds to be activated in a given department of a given hospital; the minimum number of departments in each hospitals; the number of beds to be activated in a new department; the number of beds to be activated in a given territory, and the departments that are compulsory in all the hospitals.
- Help (Help in Figure 7.2): containing a description all the system functionalities.

From the localisation section, starting from the actual configuration of a given healthcare service network (Initial Scenario), the user can make a "what-if analysis" related to the specific scenarios, described in the following.

- Closure of hospitals (Chiusura presidio in Figure 7.3): the user can select one or more hospitals to be closed and analyse the effects of such event in terms of total satisfied demand, of satisfied demand for each of the departments of the hospitals in the new network and of the degree of utilisation of beds in each of the department of the new hospital configuration.
- Closure of one or more departments (Chiusura reparto in Figure 7.3): the user can select one of more departments to be closed and analyse the effects of such event in terms of the total satisfied demand; the demand satisfied for each of the departments of the hospitals in the new network and the degree of utilisation of beds in each of the department of the new hospital configuration.
- Variation in the demand of healthcare services (Variazione domanda in Figure 7.3): the user can select the expected increase in the demand of beds for one or more departments and can study the impact of such choice on the utilization degree of the beds in all the departments of the hospital network.

The Initial scenario page (see in Figure 7.2) reports a map with all the hospitals that are active (on the basis of the data in the database) in a certain period of time (Periodo di riferimento in Figure 7.2), chosen by the user.

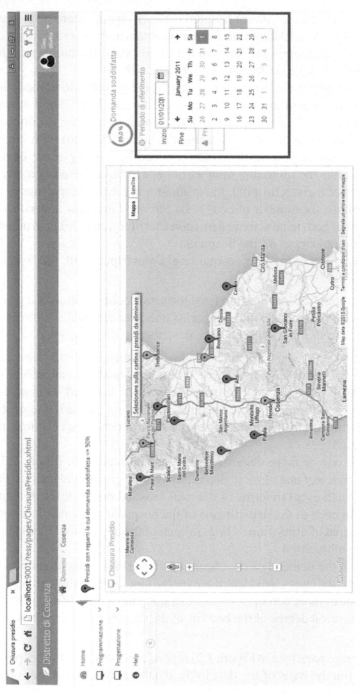

Figure 7.2   Initial network configuration.

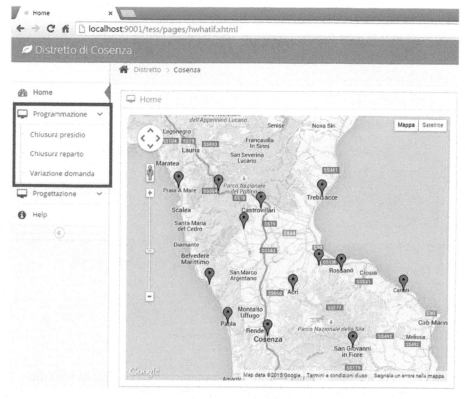

Figure 7.3 Localization section.

In particular, in Figure 7.2, the healthcare network configuration, provided by the local healthcare authority "Azienda Sanitaria Provinciale (ASP)" of Cosenza, a city in the region of Calabria, is depicted. The data related to the year 2011 and the network configuration involving the hospitals located at Acri, Castrovillari, Cetraro, Corigliano, Lungro, Mormanno, Paola, Rossano, Trebisacce, San Giovanni in Fiore, Cariati, and Cosenza are taken as initial network configuration. The hospitals in red on the map are those with a demand equal to or lower than the 50% of the actual capacity.

Information regarding a specific hospital (i.e., the total number of the active departments (n.reparti in Figure 7.4) and corresponding list [Obstetrics and Gynaecology (OSTETRICIA E GINECOLOGIA); Paediatrics and Neonatology (PEDIATRIA E NEONATOLOGIA); Urology (UROLOGIA); Nursery (NIDO)], and the percentage of satisfied demand for each department can be visualised by clicking on it (Figure 7.4).

Figure 7.4 Hospital characteristics.

Starting from the initial configuration (left menu), it is possible to start the "what-if analysis" by clicking on one of the possible alternatives.

Let's assume that the user wants to evaluate the specific scenario in which one or more hospitals are closed. By clicking on "Chiusura presidio", the page reported in Figure 7.5 will be opened, where the hospitals where some of the departments are under-utilised (the beds utilisation is under the 50%), are depicted in green while on the top right the total percentage of satisfied demand (Domanda soddisfatta) is reported (97.0%) in our example.

To evaluate the configuration in which one or more hospitals are closed, it is sufficient to click on the corresponding hospital icons. Indeed, by clicking first on the hospitals to be closed (i.e., Corigliano Calabro and San Giovanni in Fiore in our example) and after that on the button "Elabora" (Elaborate) the results will be displayed on a new page.

In particular, in Figure 7.6 the last and new configurations are compared in terms of total satisfied demands and demand satisfied for each department of the hospitals, available in the new healthcare service network.

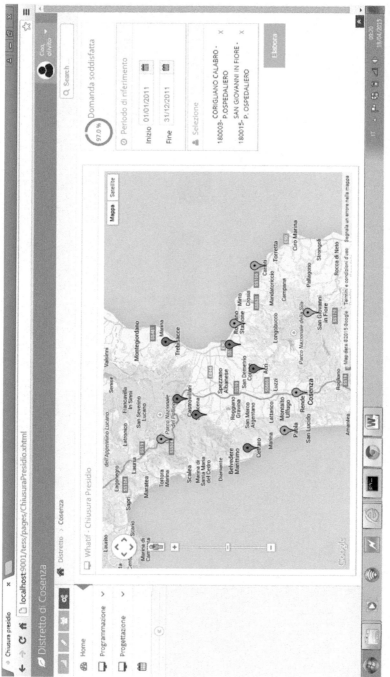

Figure 7.5 Hospital closure.

Figure 7.6   Hospital closure: The new configuration.

It is also possible to compare the initial (Pre elaborazione in Figures 7.6–7.8) and the post elaboration (Post elaborazione in Figures 7.6–7.8) scenarios in terms of % Satisfied Demand for each of the departments (Figure 7.7), % Bed Utilisation, and % Department Utilisation.

For example, Figure 7.8 shows the new situation in terms of percentage of bed utilisation after the closure of an hospital.

From the reorganisation section, a "what-if analysis" can be performed. More specifically, the effects related to a variation in the supply of healthcare services can be analysed.

The reorganisation home page shows the initial configuration (Figure 7.9), obtained by setting in an appropriate way the parameters of the reorganization models, embedded into the DSS.

Starting from the initial configuration, the following parameters can be changed:

1. number of beds available in the territory of interest;
2. number of beds to be activated in each of the department of the area of interest;
3. minimum numbers of beds to be activated in each of the departments;
4. minimum number of departments to be activated when a hospital is opened;
5. type of departments that must be open whenever a hospital is activated;
6. maximum number of beds to be activated in departments not currently present in a specific hospital.

Once, one or more parameters are changed, a new simulation can be run and a comparison between the Initial Scenario and the Post-Elaboration Scenario can be obtained, in terms of hospitals with active departments and number of beds for each department (Figure 7.10).

On the basis of the considerations reported above, it is evident that the developed "what-if analysis" is an important tool for the decision maker when variations on some fundamental parameters take place.

In particular, by evaluating the impact of these variations on the performance of the healthcare service network, it is possible to take decisions having in mind their effect and adopting solutions guaranteeing an appropriate level of service.

7.5 The "Expert Users" Functionalities of the DSS

We describe some of the functionalities of the DSS for addressing inventory management problems in healthcare.

Figure 7.7   Satisfied demand (%).

Figure 7.8 Bed utilisation (%).

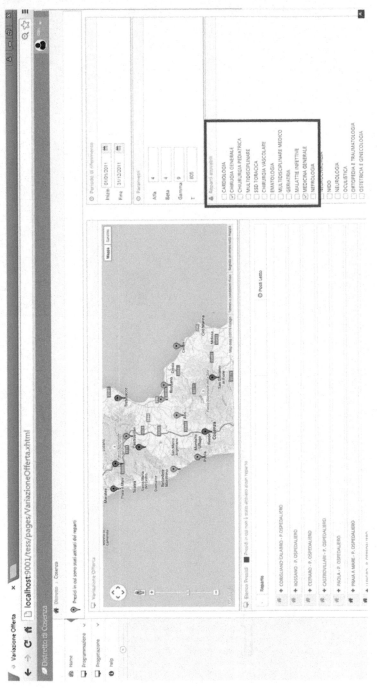

Figure 7.9   Network reorganisation.

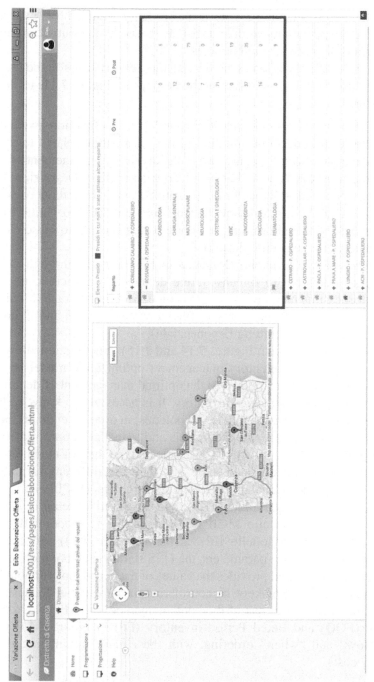

Figure 7.10 Network reorganisation: Output.

In particular, in this section we discuss the modality designed for "expert users", that give the possibility to take decisions, based by the output of the developed optimisation models and policies.

Once the user authentication is made and the "expert users" modality is selected, the user can access, from the home page in Figure 7.11, to the following features:

- Data Management (Gestione Dati in Figure 7.12): this function give the possibilities to introduce and modify all the information related to the drugs and the pharmacies that are part of the inventory management system like the drug code (Codice Farmaco); drug description (Descrizione); active ingredient code (Codice Principio Attivo); active ingredient typology (Tipologia); VAT rate (aliquota Iva); unit cost (Costo Unitario); suppplier code (Codice Fornitore); unit price (Prezzo Unitario); discount rate (Sconto); discount type (Forma).

- Demand Forecast (Previsione Domanda in Figure 7.13): through this function, it is possible to forecast future demand. In particular, three different forecasting methods are implemented: Moving Average (Media Mobile), Weighted Moving Average (Media Mobile Pesata) and Exponential Smoothing (Smoothing Esponenziale).

- Simulazione (Simulation in Figures 7.14 and 7.15): this function allows the users to apply the inventory management optimization models and policies, to take decisions related to the optimal management of departments and hospital pharmacies inventory. It is necessary to know well the embedded optimization models and policies, in order to enter all the parameters required to execute the models and to manage appropriately the obtained outputs. In particular, once selected, on the left menu, one of the models introduced in Section 7.3 (they are different variants of the Multi-product Capacitated Inventory Problem (MCIP) and of the Integrated Inventory Problem (IIP)) and discussed in detail in [33] (Modello MCIP 1; Modello MCIP 2; Modello IIP 1, Modello IIP 1 delta; Modello IIP 2; Modello IIP 2 delta in Figure 7.14) and once inserted all the necessary parameters, the DSS starts the elaboration and computes the models and provides the values of the dependent variables. Moreover, the decision makers, by considering the demand as stochastic and stationary, can select one of the implemented policies (Fixed Order Quantity (FOQ) and Fixed Period inventory (FP) in Figure 7.15) to know "how" and "when" ordering, with the aim of minimizing the inventory costs.

Figure 7.11   Home inventory management.

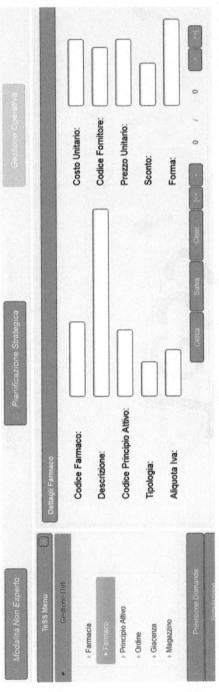

Figure 7.12   Data management.

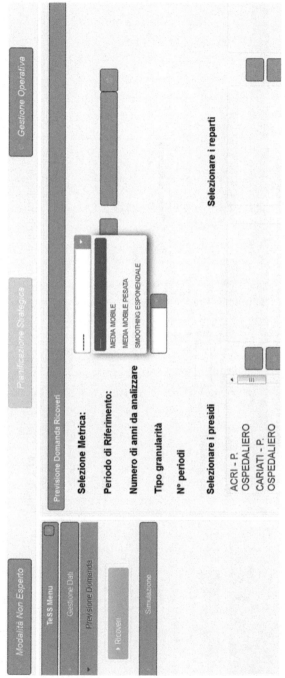

Figure 7.13 Demand forecast.

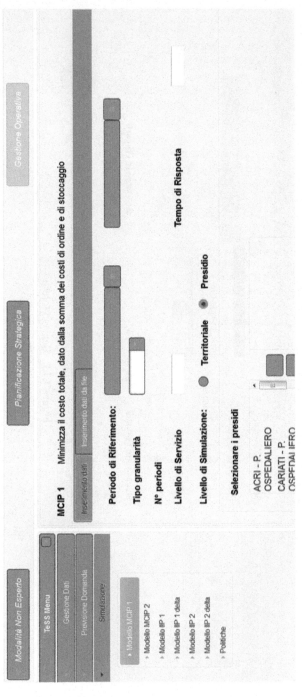

Figure 7.14 Optimisation models selection and parameter introduction.

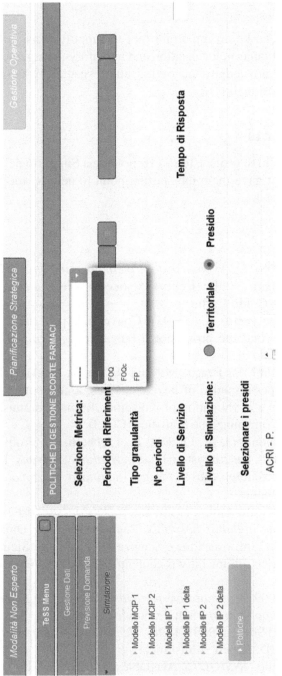

Figure 7.15   Policy selection and parameter introduction.

7.6 Conclusions

The described DSS can be an important tool to support decision maker in healthcare and in particular in localisation and inventory management issues. Its use requires some knowledge about optimisation especially when exploring its "expert users" functionalities.

Acknowledgements

The authors are grateful to project TeSS (Tecnologie a Supporto della Sanità-PON04a3 00424) that gave them the starting point to address the problems discussed in this chapter.

References

[1] Bachelet, V. (2008). *Libro bianco sui principi fondamentali del Servizio Sanitario Nazionale.* Italy: Centro di ricerca sulle amministrazioni pubbliche, Luiss Guido Carli.

[2] Lega, F., and De Pietro, C. (2005). Converging patterns in hospital organization: beyond the professional bureaucracy. *Health Policy* 74, 261–281.

[3] Rodella, S., Bellini, P., Braga, M., and Rebba, V. (2003). Measuring and comparing performance of health services: a conceptual model to support selection and validation of indicators. Roma: Commissione per la Garanzia dell'Informazione Statistica (CGIS).

[4] Lo Scalzo, A., Donatini, A., Orzella, L., Cicchetti, A., Profili, S., and Maresso, A. (2009). Health system review. *Health Syst. Trans.* 11, 1–216.

[5] Ministero della Salute. (2011). *Piano Sanitario Nazionale.* Rome: Ministero della Salute.

[6] Murgia, G., and Battistoni, E. (2012). *A decision support system for strategic planning in public hospitals.* Rapporto Tecnico, Dipartimento di Ingegneria dellâ Informazione e Scienze Matematiche (Siena). Available at: http://gerico.dii.unisi.it/wp-content/uploads/2012/06/Murgia.pdf 2012.

[7] Ozcan, Y. A. (2009). *Quantitative methods in healt care management: techniques and applications.* San Francisco: Jossey Bass.

[8] Tan, J. K. H., and Shepes, S. B. (1998). *Health decision support systems.* Silver Spring, MD: Aspen Publishing.

[9] Healthcare Insights. *INSIGHTS.* Available at: http://www.hcillc.com/

[10] *Qlik platform*. Available at: http://www.qlik.com/us/healthcare

[11] *Terso Solutions*. Available at: http://www.tersosolutions.com/hospital-inventory-management/

[12] Farahani, R. Z. (2009). *Facility Location: Concepts, Models, Algorithms and Case Studies*. Heidelberg, Germany: Physica-Verlag.

[13] Marianov, V., and Eiselt, H. A. (2011). *Foundations of Location Analysis, International Series in Operations Research and Management Science*. Berlin: Springer.

[14] Klose, A., and Drexl, A. (2004). Facility location models for distribution system design. *Eur. J. Operat. Res.* 162, 4–29.

[15] Hakimil, S. (1964). Optimum locations of switching centres and the absolute centres and medians of a graph. *Operat. Res.* 12, 450–459.

[16] ReVelle, C. S. and Swaim, R. W. (1970). Central facilities location. *Geogr Anal.* 2, 30–42.

[17] ReVelle, C., Toregas, C., and Falkson, L. (1976). Applications of the location set covering problem. *Geogr. Anal.* 8, 65–76.

[18] White, J., and Case, K. (1974). On covering problems and the central facility location problem. *Geogr. Anal.* 6, 281–293.

[19] Berman, O., and Drezner, Z. (2008). The p-median problem under uncertainty. *Eur. J. Operat. Res.* 189, 19–30.

[20] Sonmez, A. D., and Lim, G. J. (2012). A decomposition approach for facility location and relocation problem with uncertain number of future facilities. *Eur. J. Operat. Res.* 218, 327–338.

[21] Wang, Q., Batta, R., Bhadury, J., and Rump, C. (2003). Budget constrained location problem with opening and closing of facilities. *Comput. Operat. Res.* 30, 2047–2069.

[22] Bijvank, M., and Vis, I. F. A. (2012). Inventory control for point-of-use locations in hospitals. *J. Operat. Res. Soc.* 63, 497–510.

[23] Frederick, B. J. (1995). The management of the supply chain for hospital pharmacies: a focus on inventory management practices. *J. Bus. Logistics* 16, 153–173.

[24] Lapierre, S. D. and Ruiz, A. B. (2007). Scheduling logistic activities to improve hospital supply systems. *Comput. Operat. Res.* 34, 624–641.

[25] Little, J., and Coughlan, B. (2008). Optimal inventory policy within hospital space constraints. *Health Care Manag. Sci.* 11, 177–183.

[26] Vila-Parrish, A., Ivy, J., and King, R., et al. (2012). Patient-based pharmaceutical inventory management: a two-stage inventory and production model for perishable products with markovian demand. *Health Syst.* 1, 69–83.

[27] Mustaffa, N. H., and Potte, A. (2009). Healthcare supply chain management in malaysia: a case study. *Supply Chain Manag. Int. J.* 14, 234–243.

[28] Nicholson, L., Vakharia, A. J., and Erenguc, S. (2004). Outsourcing inventory management decisions in healthcare: models and application. *Eur. J. Operat. Res.* 154, 271–290.

[29] Ghiani, G., Laporte, G., and Musmanno, R. (2004). *Introduction to Logistics Systems Planning and Control.* New York: John Wiley & Sons.

[30] Waters, C. D. J. (1992). *Inventory Control and Management.* New York: John Wiley & Sons.

[31] Johnson, L. A., and Montgomery, D. (1974). *Operations Research in Production Planning, Scheduling, and Inventory Control.* New York: John Wiley & Sons.

[32] Guerriero, F., Miglionico, G., and Olivito, F. (2016). Location and reorganization problems: the calabrian health care system case. *Eur. J. Operat. Res.* 250, 939–954.

[33] Guerriero, F., Miglionico, G., and Olivito, F. (2016). *Inventory management strategies for the calabrian hospitals system.* Techical Report DIMEG, Universitá della Calabria, Padova.

8

Deep Learning Classifier for Fall Detection Based on IR Distance Sensor Data

Stanisław Jankowski[1], Zbigniew Szymański[1], Uladzimir Dziomin[2], Paweł Mazurek[1] and Jakub Wagner[1]

[1]Institute of Radioelectronics and Multimedia Technologies,
Faculty of Electronics and Information Technology, Warsaw University
of Technology, ul. Nowowiejska 15/19, 00-665 Warsaw, Poland
[2]Brest State Technical University, Department of Intelligent Information
Technology, Moskovskaja str. 267, 224017 Brest, Belarus

Abstract

The goal of research is the fall detection in elderly residents based on infrared depth sensor measurements. We present the methodology of data acquisition, preprocessing, and the feature extraction. Our attention is focused on statistical properties as generalisation. We studied three classification scenarios – neural network, neural network enhanced by feature selection, and deep learning system. The effectiveness of discriminative statistical classifiers (multilayer perceptron) of the deep learning system is improved by addition of feature selection block by Gram-Schmidt orthogonalisation, which determines the ranking of the features, and NPCA block, which transforms the raw data into a non-linear manifold and reduces the dimensionality of the data. Performance of our system measured in terms of sensitivity is 94% and precision is 96%, which means it can be used for real life applications.

Keywords: Fall detection, Infrared distance sensor, NPCA, Feature selection, Gram–Schmidt orthogonalisation.

8.1 Introduction

The goal of the research is the fall detection in elderly residents based on infrared (IR) depth sensor measurements. Fall detection is widely addressed

169

in recent research papers. The fall accidents of elderly people cause social and economic problems [1]. Therefore, a reliable method of fall detection, which does not violate the privacy of monitored persons, is of great importance.

The main approaches to fall detection are vision based methods (e.g., inactivity detection, body shape change analysis, and three-dimensional (3D) head motion analysis), environmental methods (e.g., IR depth sensors [2] and pressure sensors), and wearable devices (usually comprise an accelerometer) [1, 3]. The fall detection is performed by algorithms based on a fixed flowchart (e.g., detection of various threshold values [4, 5]) or by machine learning methods [6]. Comparison of the detection results is very difficult due to different study protocols, experimental scenarios, and calculated parameters presenting the effectiveness of developed methods. The sensitivity of methods based on depth sensors vary from 83% (no machine learning) [7] up to 91% (machine learning approach) [6]. Our research explores the data obtained from the IR depth sensors.

This chapter presents methodology applied to data derived from people falling. The data and the scenarios are described in Jankowski et al. [8]. Our attention is focused on statistical properties, as generalisation, i.e., the prediction of successful classification (fall detection) on the general population.

We plan to perform the following steps:

- Feature extraction from the trajectories described in Section 8.3.
- Input features selection by Gram–Schmidt orthogonalisation [9, 10].
- Projection of input features vectors after Gram–Schmidt selection onto non-linear manifold by the non-linear principal component analysis (NPCA) [11].
- Learning of multilayer perceptron.
- Tests of classifiers, e.g., by cross-validation method.
- The selection of the best classifier for real-life application.

8.2 Statistical Classification

We decided to use the multilayer perceptron as the statistical learning classifier for fall detection, because the recognition based on deterministic approach is not known. The goal of the supervised learning is to assign to vectors from a given input space $X \subseteq \mathrm{R}^n$ values from the output domain Y based on a training set. For two-class classification $Y = \{-1, 1\}$, where a fall is labelled as 1, a non-fall as -1. These labels are assigned by an expert, hence the term supervised learning. A training set S is a collection of training examples defined as:

$$S = ((\mathbf{x}_1, y_1), (\mathbf{x}_2, y_2), \ldots, (\mathbf{x}_N, y_N)) \subseteq (X \times Y)^N. \qquad (8.1)$$

We denote by: N – the number of examples, x, examples and y, their labels.

As the learning system is obtained based on experimental results without prior knowledge our preliminary considerations are general. The successful classification can be reached if the fundamental relations of three integers are kept:

- n – the dimension of the input space equal to the number of descriptors;
- N – number of training examples;
- q – number of adjustable system parameters (weights) that determines the system complexity.

The number of training examples must satisfy the inequality [12]:

$$N > 2(n + 1) \qquad (8.2)$$

If a 2-class training set does not satisfy (Equation (8.2)), any random hyperplane separates classes without error with the probability 1. Hence, the statistical classifier is useless.

The generalisation can be estimated by testing the learning system on the data set not explored during learning – the test set. If the data set is limited the generalisation can be estimated by cross-validation. The generalisation error bounds are functions of the q/N ratio. The complexity increase enables to decrease the training set error due to overfitting to training data, while the test set error increases, so the generalisation is lost. The overfitting can be avoided by imposing the limitation on the number of the adjustable system parameters. Therefore, in order to improve the classifier generalisation, we attempt to decrease the dimension of the input space n by removing irrelevant and redundant features. The number of hidden neurons depends on the shape of discriminant surface between two classes. Hence, the appropriate input data representation may cause also the decrease of the neural network complexity – lower number of hidden neurons. The overfitting can be controlled by using cross validation.

There are two aspects of statistical learning systems:

- learning algorithm;
- statistical properties.

8.2.1 Selection of Statistical Learning Algorithm

Statistical learning systems are wide area of research in artificial intelligence. There are many well-defined models. It can be proved that the best classification limit, as the number of learning examples tends to infinity, is the Bayes conditional probability [13]. However, the acceptable estimation of prior conditional probability densities requires the number of examples proportional to 10^{n+1}, where n is the dimension of the input space. Therefore, the discriminative classifiers should be used. The principle of these models is to estimate the hypersurfaces that divide the input space into subspaces containing the examples (points) of considered classes.

These discriminative classifiers are based on solid mathematical theories and the successful practical applications. The most important models are: multilayer perceptron [14], support vector machine (SVM) [15, 16], least-squares support vector machine (LS-SVM) [17, 18], and relevance vector machine (RVM) [19].

These models implement various approaches to system design. The multilayer perceptron learning goal is to obtain the minimum of the least-squares error between the calculated and the desired value at the output neuron. The goal is obtained by using the gradient optimisation methods: error backpropagation, Levenberg–Marquardt method, etc.

The SVM was formulated for linearly separable classes as the maximum margin solution. This model explores the optimisation with inequality constraints. It can be proved that the separating hyperplane is defined by a subset of learning examples with positive Lagrange coefficient. However, the generalisation to non-linearly separable and non-separable case can be obtained by introducing special kernel functions defined by specific parameters and regularisation coefficient. These additional so-called hyperparameters are optimised by various usually heuristic algorithms. Therefore, the SVM solution is not unique and the generalisation is hard to estimate.

In practice, the results obtained by modern statistical classifiers are similar. We decided to start our research in fall recognition project by using the multilayer perceptron due to efficient learning methods as well as clear analytical, geometric, and statistical interpretation that can be used to optimise its generalisation score.

8.2.2 Multilayer Perceptron as Discriminative Classifier

Multilayer perceptron is widely used neural network [20, 21]. This is a network built of three layers of neurons, the layers are fully connected and the connections are presented by real numbers adjusted during learning phase. The hidden and the output neurons are non-linear processing units with activation function, e.g., a hyperbolic tangent. The network is defined as follows:

- Network inputs are the descriptors of classified objects.
- The weights of the hidden layer are directional coefficients of separating hyperplanes and the separating hypersurface is a linear combination of these hyperplanes. Hypersurface separating the class C_i from the other classes is a hypersurface $g_i(x) = 0$, where g_i is i-th output.
- Network outputs: for two-class problem only one output neuron is used with hyperbolic tangent activation function: its saturation values are 1 and -1, that can correspond to class labels. During learning stage, for an object of class C_i, the desired i-th output is $+1$, otherwise is -1.

Hence, the separating hypersurface is a superposition performed by the output neuron of hyperplanes represented at the input of the hidden neurons This fact enables direct interpretation of the appropriate structure of a multilayer perceptron. The weights are modified in order to satisfy the minimum of the cost function. Usually it is a sum of residual errors at the network output. The learning/training is performed by gradient methods. It can be stated that multilayer perceptron training is equivalent to non-linear least squares problem. It is proved that one hidden layer is sufficient to solve any problem of regression or classification.

8.2.3 Preprocessing and Variable Selection

The first step in any model design procedure consists of reducing the dimensionality of the input space. Our aim is to find the intrinsic dimension of the input vector as small as possible and to find a more compact input representation, while preserving the amount of relevant information by using NPCA. The second problem is to select the inputs relevant to the modelling of the quantity of interest by using the Gram–Schmidt orthogonalisation.

NPCA (described in Section 8.4.5) is a procedure that is carried out in *representation space*, in which each observation is represented by a point, whose coordinates are the values of the factors that correspond to that observation. The Gram–Schmidt orthogonalisation for the selection of inputs is carried out in *the observation space*, where each factor is represented by a vector, the components of which are observations of this factor in the database. The dimension of representation space is the number of factors of the model and the dimension of observation space is the number of observations in the database.

The preprocessing can reduce the number of inputs and thus to decrease the classifier complexity (number of calculated parameters) and finally the generalisation ability.

8.2.4 Generalisation and Quality Prediction

The practical goal of statistical learning systems is the balance of the training error and test error. High-variance learning methods represent the training set well, but tend to overfit to noisy data. Methods with high bias typically generate more simple models that don't tend to overfit. Overfitting can be estimated by the cross-validation method [22, 23].

The model overfitting is a property of the function:

- that is able to fit exactly to all data of the training set. Its parameters are used to represent training data even if noise is added,
- whose complexity expressed as the number of degrees of freedom is high enough to fit to any element of the training set even with additive noise.

This effect is local: in some intervals of the input variables the function uses some degrees of freedom (network weights) to fit the function exactly to some desired values.

Leave one out cross-validation was performed on the data set to evaluate presented learning system. In order to validate the classifiers false positive (FP), false negative (FN), true negative (TN) and true positive (TP), precision, and sensitivity parameters were calculated. Precision is defined as

$$\text{precision} = \frac{\text{TP}}{\text{TP} + \text{FP}} \times 100\%. \tag{8.3}$$

The sensitivity is defined as

$$\text{sensitivity} = \frac{\text{TP}}{\text{TP} + \text{FN}} \times 100\%. \tag{8.4}$$

Precision score of 100% for binary classification means that every item labelled as belonging to positive class does indeed belong to positive class. Sensitivity of 100% means that every item from positive class was labelled as belonging to positive class.

8.3 Methodology of Data Generation

8.3.1 Data Acquisition

The data for testing fall detection algorithms has been acquired by means of two synchronised IR depth sensors, called sensors S_1 and S_2 hereinafter [24]. The sensors being parts of two Kinect devices (model 1473) have been used for experimentation. Their configuration, relative to the observed area, is

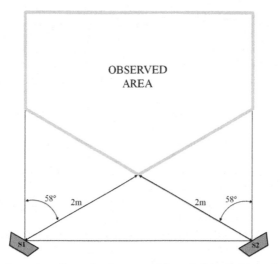

Figure 8.1   Configuration of two depth sensors (S_1 and S_2) relative to the observed area.

presented in Figure 8.1. The distance between them was set to *ca.* 3 m. A monitored person moved at the distance of *ca.* 1.5–5 m from each of them. Each experiment lasted 10 s, and consisted in recording data from both sensors simultaneously, with the frame frequency of 30 fps. It resulted in a sequence of 300 depth images, i.e., 480×640 matrices of integer numbers representative of the distance from the sensors.

A set of 18 fall scenarios and 18 scenarios corresponding to other actions of a monitored person, called *non-falls*, has been designed. Short descriptions of those scenarios are provided in Table 8.1. All the scenarios have been realised by two actors. The whole set of acquired data contains 144 sequences of depth images since each out of 36 scenarios has been repeated by two actors, and recorded by two sensors S_1 and S_2.

8.3.2 Data Preprocessing

Each pixel of an image, mapped by means of a sensor S_1 or S_2, is represented by a triplet of integer numbers (i, k, d), where $i \in \{1, 2, \ldots, I\}$ is the column index, $k \in \{1, 2, \ldots, K\}$ is the row number, and $d \in \{1, 2, \ldots, 5000\}$ is the distance to the sensor (in millimetres). Since such a relative representation of the image may be a source of ambiguity – the same person may appear as larger or smaller, depending on the distance from a sensor – it has been transformed into the absolute representation based on global space coordinates (x, y, z).

Table 8.1 Experiment scenarios

Scenario Name	Description
Fall 1–12	Entering the observed area from various directions followed by a forward fall ending with lying flat
Fall 13	Standing still in the observed area followed by a backward fall ending with lying flat
Fall 14	Standing still in the observed area followed by a lateral fall to the right ending with lying flat
Fall 15	Standing still in the observed area followed by a forward fall ending with lying flat
Fall 16	Standing still in the observed area followed by alateral fall to the left ending with lying flat
Fall 17	A backward fall with rotation following an attempt to sit down, ending with lying flat
Fall 18	A forward fall with rotation during bending down, ending with lying flat
Non-fall 1	Sitting down on a chair
Non-fall 2	Sitting down on the floor
Non-fall 3	Bending down and rising up
Non-fall 4	Lying down on the floor
Non-fall 5	Croaching
Non-fall 6	Walking around the observed area
Non-fall 7–12	The same as non-fall 1–6, in smaller distance from the sensors
Non-fall 13–17	The same as non-fall 1–5, after entering into the observed area from outside
Non-fall 18	Entering into the observed area, stopping there, and walking outside of it

The corresponding mathematical procedure, whose detailed description may be found in another paper presented at this conference [24], consists of two operations:

- identification of a set P of pairs (i, k), representative of the silhouette of a monitored person (Figure 8.2);
- transformation of triplets (i, k, d), corresponding to P, into triplets (x, y, z).

The absolute coordinates $(x_{i,k}, y_{i,k}, z_{i,k})$, calculated for all $(i, k) \in P$, have been used for computation of the coordinates of the "mass centre" of the silhouette:

$$(x_C, y_C, z_C) = \frac{1}{|P|} \sum_{(i,k)\in P} (x_{i,k}, y_{i,k}, z_{i,k}), \qquad (8.5)$$

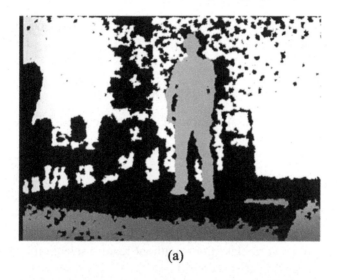

(a)

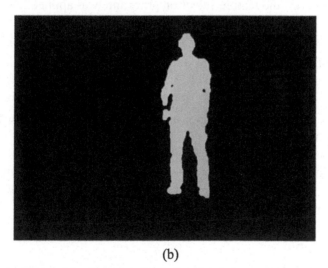

(b)

Figure 8.2 Extraction of the silhouette of a monitored person: (a) the original image recorded by the sensor S_1; (b) the result of extraction.

and its magnitude, i.e., the effective reflection area:

$$M = \frac{1}{|\mathrm{P}|} \left(\sum_{(i,k)\in\mathrm{P}} d_{i,k} \right)^2. \tag{8.6}$$

The application of the above-described procedure to a sequence of depth images, acquired at the selected time instants $t_n (n = 1, 2, \ldots,)$, resulted in four-dimensional trajectories:

$$\{x_C(t_n), \ y_C(t_n), \ z_C(t_n), \ M(t_n)\}, \tag{8.7}$$

which may be used for classification of events recorded by the sensor S_1 or S_2.

8.4 Deep Learning Classifier

We studied the effect of various classification scenarios on the final score. The three considered scenarios were:

- regular neural network classifier – without feature selection and NPCA used as reference for other scenarios,
- neural network – the feature selection procedure was applied to the input data,
- deep learning system [25–30] based on feature selection and NPCA.

The structure of proposed learning systems is shown in Figure 8.3. The effectiveness of discriminative statistical classifiers (in our case the neural network) is improved by addition of blocks performing the data pre-processing:

- the feature selection block (by Gram–Schmidt orthogonalisation) determines the ranking of the features – only three most significant features are used in the next stage, and
- NPCA block – transforms the raw data into a non-linear manifold. The NPCA procedure reduces the dimensionality of the data to two dimensions.

8.4.1 The Data Set

The creation of the data set used for classification purposes included following steps:

- Data filtration: the data acquired from the depth sensors described by formula (8.7) was filtered using neural network regression,
- Feature extraction: Six features were computed for every recording. Each fall and non-fall motion is described by two feature vectors corresponding to recordings from two depth sensors. The data set consists of 144 feature vectors including six elements.

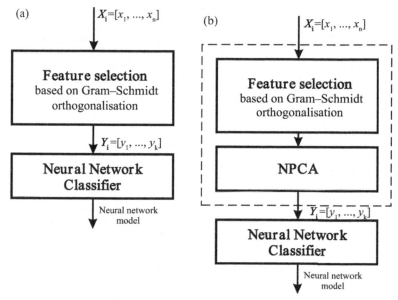

Figure 8.3 The structure of (a) neural network classifier and (b) the deep learning system.

- Feature selection: a subset of features was selected in order to maintain high-classification performance.

8.4.2 Data Filtration

The experiments revealed the need for data filtration. It was performed by neural network regression. The X, Y, Z position of mass centre and the magnitude were filtered. X, Y, Z velocities and X, Y, Z accelerations were calculated from the filtered position variables. The results of the filtering are presented in Figures 8.4 and 8.5. Network architecture used for filtration (regression task):

- number of inputs: 1 (sample number)
- hidden layer: 20 neurons (transfer function: hyperbolic tangent)
- output layer: 1 neuron (transfer function: linear)
- learning algorithm: Levenberg–Marquardt backpropagation
- maximum no of epochs: 1200 (sufficient to obtain the minimum of the goal function)

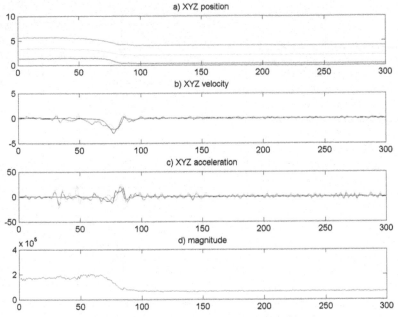

Figure 8.4   Example of original data before filtration.

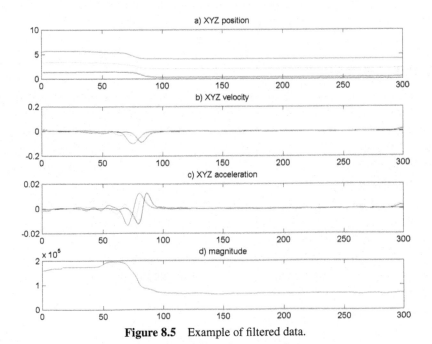

Figure 8.5   Example of filtered data.

8.4.3 Feature Extraction

For each recording six features were computed:

- maximum velocity in the *XY* plane;
- maximum absolute value of velocity along *Z* axis;
- maximum acceleration in the *XY* plane;
- maximum absolute value of acceleration along *Z* axis;
- magnitude for 10th sample after sample where absolute value of z acceleration reaches its maximum;
- z coordinate for 10th sample after sample where absolute value of z acceleration reaches its maximum;

Figure 8.6 shows extracted features from the recording of the fall scenario 1 – the feature values are marked by the dots on the graphs. Figure 8.7 shows extracted features from the recording of the non-fall scenario 1. While the shapes are similar, the graphs differ in values.

8.4.4 Feature Selection

The orthogonalisation procedures enable us examination of the influence of every input feature on the output vector. The presented method uses simple ranking formula and the well-known Gram–Schmidt orthogonalisation

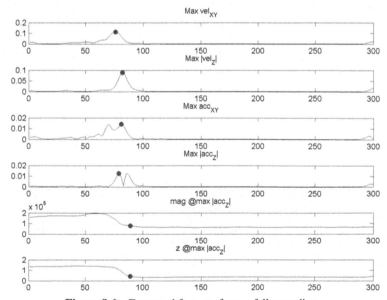

Figure 8.6 Extracted features from a fall recording.

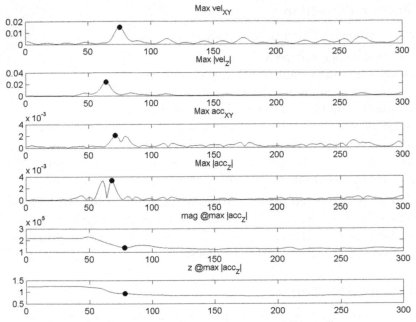

Figure 8.7 Extracted features from a non-fall recording.

procedure for pointing out most salient variables of the model [25, 26]. The data set consists of N input–output pairs containing the candidate features (input) and the measurements of the process to be modelled (output).

We denote by: Q – number of candidate features, N – number of measurements of the process to be modelled, $x^i = [x^i_1, x^i_2, \ldots, x^i_N,]$ – the vector of the i-th feature values of N measurements, y_p – the N-dimensional vector of the classifier target values. We consider the N by Q matrix $X = [x^1, x^2, \ldots, x^Q,]$. The ranking procedure starts with calculating correlation coefficient.

$$\cos^2(x^k, y_p) = \frac{(x^k \cdot y_p)^2}{(\|x^k\|^2 \cdot \|y_p\|^2)}. \tag{8.8}$$

The larger it is, the better the k-th feature vector explains the y_p variation. As the first base vector, we pick the one with the largest value of correlation coefficient. All the remaining candidate features and the output vector are projected onto the null subspace (of dimension $N - 1$) of the selected feature. Next, we calculate correlation coefficient for the projected vectors and again pick the one with the largest value of this quantity.

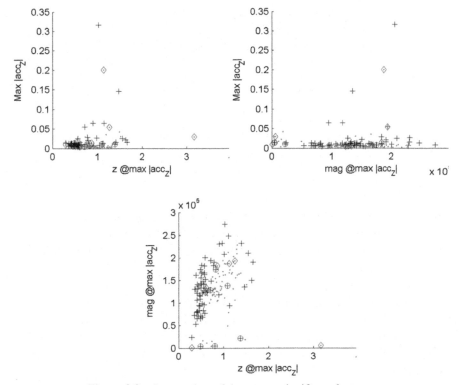

Figure 8.8 Scatter plots of three most significant features.

The remaining feature vectors are projected onto the null subspace of the first two ranked vectors by the classical Gram–Schmidt orthogonalisation. This procedure is continued until all the vectors x^k are ranked.

Scatter plots of three most significant features are shown in Figure 8.8. Data points representing falls are marked by + signs and non-falls are marked by dots.

Three most relevant features were selected:

- coordinate for 10th sample after sample where absolute value of z acceleration reaches its maximum,
- magnitude for 10th sample after sample where absolute value of z acceleration reaches its maximum, and
- maximum absolute value of acceleration along Z-axis.

Table 8.2 presents three most significant features and the calculated correlation coefficients according to the formula (8.8). In order to reject the irrelevant

Table 8.2   Experiment scenarios

Rank	Feature Name	$\cos^2 t(x^k, y_p)$
1	z @ max $\mid \mathrm{acc}_Z \mid$	0.203651
2	mag @ max $\mid \mathrm{acc}_Z \mid$	0.162742
3	max $\mid \mathrm{acc}_Z \mid$	0.136410
–	random feature	0.001144

inputs, we compare the correlation coefficients with the rank of a random probe (presented in last row of Table 8.2).

8.4.5 NPCA

Principal component analysis (PCA) is a statistical procedure that uses an orthogonal transformation to convert a set of observations of possibly correlated variables into a set of values of linearly uncorrelated variables called principal components.

As the projection subspace is unknown, the calculation is made in a system called autoencoder. The projection subspace is represented by a bottleneck layer. The vector of the training set is given at the input and at the output. The weights in the input and the output layers are set to be equal: $W = W'$. The applied iterative algorithm attempts to the eigenvectors of the covariance matrix defined as

$$\Sigma = \sum_i \left(\mathbf{x}^i - \bar{\mathbf{x}}\right)\left(\mathbf{x}^i - \bar{\mathbf{x}}\right)^{\mathrm{T}} \tag{8.9}$$

$$\bar{\mathbf{x}} = \frac{1}{N} \sum_1^N \mathbf{x}^i. \tag{8.10}$$

The eigenvalues represent the relevance of a given new coordinate. The error between the input and the output vector is the information loss.

NPCA is a non-linear generalisation of standard principal component analysis (PCA) [29–31].

The NPCA transforms the raw data into a non-linear manifold spanned by the eigenvectors of the correlation matrix [29–31]. NPCA neural network is a two layer neural network. The added layers are composed of units with non-linear activation function (e.g., hyperbolic tangent). The learning explores Oja's rule [32]. Hence, the system is named "deep learning system". The layers are interconnected by bidirectional links. Each layer can be considered as input or an output. If the layer works as an input, it performs the distribution function. In other case, it performs processing function. The coefficients for forward and backward connections are represented as W and W'. The computation explores only the correlation between pairs of input vectors without consideration of the class label. This is an unsupervised learning method. The architecture of NPCA neural networks are illustrated in Figures 8.9 and 8.10.

The NPCA neural networks (NN) can be used for data compression and for restoration of the already compressed data. Data compression was

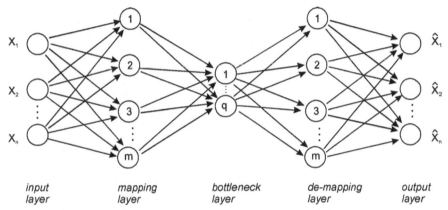

Figure 8.9 The architecture of NPCA neural network (learning stage).

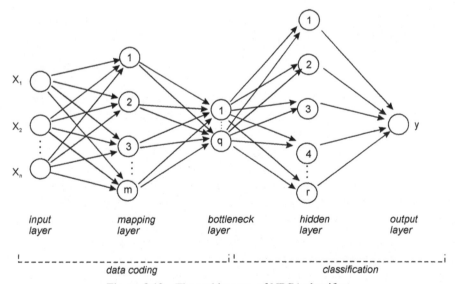

Figure 8.10 The architecture of NPCA classifier.

carried out by direct transformation of information in accordance to the expression:

$$Y = F(S^j), \tag{8.11}$$
$$S^j = W'X + T^j. \tag{8.12}$$

Restoration or reconstruction of data occurs when converting information:

$$\overline{X} = F(S^i) \tag{8.13}$$
$$S^i = WY + T^i. \tag{8.14}$$

Function F is an activation function of neural elements. It can be either linear or non-linear functions. By using a linear activation function, the previous equations take the form:

$$Y = WX + T^j \tag{8.15}$$
$$\overline{X} = WY + T^i. \tag{8.16}$$

The weight matrix can be acquired by the method of principal components. In this case, the columns of the matrix W' are equal to the eigenvectors of the covariance matrix. Then:

$$W' \equiv W. \tag{8.17}$$

Thus, the weight coefficients of the PCA neural network are determined using the method of principal components. In this case, W is an orthogonal matrix, and

$$WW^T = 1 \tag{8.18}$$

The network which weights are determined by the method of principal components is called the PCA network [11]. We consider the methods of training neural networks recirculation.

In our learning system, NPCA procedure accepts at the input the data set containing three features determined by the feature selection procedure. The final goal of NPCA block is to decrease the dimensionality of the data used for classification. The best NPCA architecture chosen for our system is 3-5-2-5-3 (Figure 8.9):

- $n = 3$ inputs (features determined by feature selection procedure),
- $m = 5$ neurons in denoising layer, which scales the data from 3D to five,
- $q = 2$ neurons in bottleneck layer, which compresses the dataset to two dimensions. The compressed data is used for classification.

The Figure 8.11 shows the manifold for the 3D data after the NPCA learning. The bottleneck principal component data (used for classification) is shown in Figure 8.12. Points belonging to positive class (falls) are denoted by + sign, points belonging to negative class (non-falls) are denoted by dots.

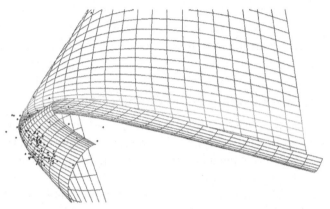

Figure 8.11 The manifold of the 3Ds dataset obtained by the NPCA (performed after the feature selection).

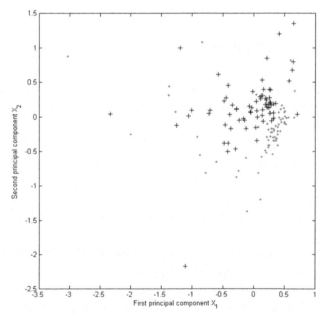

Figure 8.12 Two-dimensional data after compressing by the NPCA procedure (performed after the feature selection).

8.5 Results

There are 36 motion scenarios. Each scenario was performed by two persons and the data acquisition was performed by two sensors. Thus, the data set comprises 144 vectors (one vector for each recording). Three learning systems were tested – regular neural network, neural network enhanced by feature selection, and deep learning systems based on NPCA. Input data set for enhanced neural network was 3D (obtained by the feature selection procedure). Input data set for deep learning system was two dimensional (obtained by the NPCA procedure).

8.5.1 Neural Network Enhanced by Feature Selection

Neural network used for classification comprised three inputs, 20 neurons in hidden layer (transfer function: hyperbolic tangent), 1 neuron in output layer (transfer function: hyperbolic tangent), the learning algorithm was Levenberg–Marquardt backpropagation [33, 34].

Neural network classification results are presented in Table 8.3. The first classification scenario (regular neural network, without additional processing block) serves as a reference. Sensitivity is 89% and precision is 86%. In the second scenario, where feature selection was applied, sensitivity value increased to 92% and precision increased to 93%.

8.5.2 Deep Learning System

Different number of neurons in denoising layer were considered. The results are presented in Table 8.4. The final number of bottleneck neurons was set to 2 after the experiments. The less and greater number of bottleneck neurons make classification result worse or at the same level.

Neural network used for classification comprised (Figure 8.10):

- $n = 3$ inputs (features determined by feature selection procedure),
- $m = 5$ neurons in denoising/mapping layer,
- $q = 2$ bottleneck neurons,
- $r = 5$ neurons in hidden layer (transfer function: hyperbolic tangent), and
- 1 neuron in the output layer (transfer function: hyperbolic tangent).

The learning algorithm was Levenberg–Marquardt backpropagation.

Classification results of our deep learning system are presented in Table 8.2. When feature selection and NPCA was applied, sensitivity value increased to 94% and precision increased to 96%.

Table 8.3 Classification results

Classification Scenario	TP	TN	FP	FN	Sensitivity (%)	Precision (%)
(1) NN	64	62	10	8	89	86
(2) Feature selection (NN)	66	67	5	6	92	93
(3) Feature selection, NPCA, NN	68	69	3	4	94	96

Table 8.4 Classification results versus size of bottleneck layer

Bottleneck Size	Error	Sensitivity	Precision
1	21.5%	87.5%	74.1%
2	18.1%	76.4%	85.9%
3	6.9%	97.2%	89.7%
4	5.6%	91.7%	97.1%
5	4.9%	94.4%	95.7%
6	6.9%	89.7%	97.2%

False negative is the most undesired outcome because it means that the classifier missed a fall of monitored person. The consequences of false positives are less serious – it is just a false alarm. The results analysis reveals that the fall scenario 12 is the most challenging one. The outcomes of three test cases based on it resulted in false negatives. Further research should be made to identify the cause of such behaviour.

The neural classifier without any enhancements yielded eight false negatives. Application of additional processing blocks in the deep learning classification enabled a decrease of this value by 50% to 4.

8.6 Conclusions

We have shown that the smart data preprocessing may reveal the influential properties of the data set so that the classification score may be improved by using the classifier of reduced complexity. We focused on two phenomena:

- the feature selection based on the correlation of the input data and the output label, and
- the reconstruction of the non-linear manifold containing the training data by NPCA.

Performance of our system measured in terms of sensitivity is 94% and precision is 96%. We believe it can be used for real-life applications although there are no fixed thresholds specifying when a system can be used in such applications. However, our results are at the state of the art level comparable to other solutions or even exceeding them as described in the section "Introduction (8.1)".

The deep learning classifier structure is much simpler: 5 hidden neurons compared to 20 hidden neurons in neural network. It can be concluded that the predicted generalisation ability improved. The obtained classification score for deep learning system shows that this is a limit of successful fall detection based on the studied experimental data.

References

[1] Abbate, S., Avvenuti, M., Corsini, P., Light, J., Vecchio, A. (2010). "Monitoring of human movements for fall detection and activities recognition in elderly care using wireless sensor network: a survey," in *Wireless Sensor Networks: Application-Centric Design*, eds. G. V. Merrett and Y. K. Tan, InTech, DOI: 10.5772/13802. Available from: http://www.intechopen.com/books/wireless-sensor-networks-applicati on-centric-design/monitoring-of-human-movements-for-fall-detection- and-activities-recognition-in-elderly-care-using-wi

[2] Gasparrini, S., Cippitelli, E., Spinsante, S., Gambi., E. (2014). A depth-based fall detection system using a kinect sensor. *Sensors (Basel)* 14, 2756–2775.

[3] Delahoz, Y. S., Labrador, M. A. (2014). Survey on fall detection and fall prevention using wearable and external sensors. *Sensors (Basel)* 14, 19806–19842.

[4] Dai, J., Bai, X., Yang, Z., Shen, Z., Xuan, D. (2010). "PerFallD: a pervasive fall detection system using mobile phones," in *Proceedings of the 8th IEEE International Conference on Pervasive Computing and Communications* (Rome: IEEE), 292–297.

[5] Kangas, M., Konttila, A., Lindgren, P., Winblad, I., Jamsa, T. (2008). Comparison of low-complexity fall detection algorithms for body attached accelerometers. *Gait Posture* 28, 285–291.

[6] Ma, X., Wang, H., Xue, B., Zhou, M., Ji, B., and Li, Y. (2014). Depth based human fall detection via shape features and improved extreme learning machine. *IEEE J. Biomed. Health Inform.* 18, 1–8. doi: 10.1109/JBHI.2014.2304357

[7] Bevilacqua, V., Nuzzolese, N., Barone, D., Pantaleo, M., Suma, M., D'Ambruoso, D., et al. (2014). "Fall Detection in indoor environment with kinect sensor," in *Proceedings of the (INISTA) IEEE International*

Symposium on Innovations in Intelligent Systems and Applications (Rome: IEEE), 319–324.

[8] Jankowski, S., Szymański, Z., Mazurek, P., Wagner, J., (2015). *Neural Network Classifier for Fall Detection Improved by Gram-Schmidt Variable Selection.* Rome: IEEE. 728–732.

[9] Guyon, I., Elisseeff, A. (2003). An introduction to variable and feature selection. *J. Mach. Learn. Res.* 3, 1157–1182.

[10] Stoppiglia, H., Dreyfus, G., Dubois, R., and Oussar, Y. (2003). Ranking a random feature for variable and feature selection. *J. Mach. Learn. Res.* 3, 1399–1414.

[11] Scholz, M., Fraunholz, M., Selbig, J. (2008). "Nonlinear principal component analysis: neural network models and applications," in *Principal Manifolds for Data Visualization and Dimension Reduction*, (Berlin: Springer), 44–67.

[12] Cover, T. M., (1965). Geometrical and statistical properties of systems of linear inequalities with applications in pattern recognition. *IEEE Trans. Electron. Comp.* 14, 326–334.

[13] Hastie, T., Tibshirani, R., Friedman, J., (2001). *The Elements of Statistical Learning – Data Mining, Inference and Prediction.* Berlin: Springer.

[14] Dreyfus, G., J. Martinez, M., Samuelides, M., Gordon, M. B., Badran, F., Thiria, S. L. (2005). *Hérault Neural Networks, Methodology and Applications.* Berlin: Springer.

[15] Vapnik, V., (2000). *The Nature of Statistical Learning Theory.* Berlin: Springer.

[16] Jankowski, S., Szymański, Z., Piątkowska-Janko, E., Oręziak. A. (2007). Improved recognition of sustained ventricular tachycardia from SAECG by support vector machine. *Anatol. J. Cardiol.* 7, 112–115.

[17] Suykens, J. A. K., and Vandewalle, J. (1999). Least squares support vector machine classifier. *Neural Process. Lett.* 9, 293–300.

[18] Suykens, J. A. K., Van Gestel, T., De Brabanter, J., De Moor, B., Vandewalle, J., (2002). *Least Squares Support Vector Machines.* Singapore: World Scientific.

[19] Tipping, M. E. (2000). "The relevance vector machine," in *Advances in Neural Information Processing Systems*, ed. S. Mateo (Burlington, MA: Morgan Kaufmann).

[20] Hornik, K., Stinchcombe, M., White, H., (1989). Multilayer feedforward networks are universal approximators. *Neural Netw.* 2, 359–366.

[21] Hornik, K. (1991). Approximation capabilities of multilayer feedforward networks. *Neural Netw.* 4, 251–257.

[22] Geman, S., Bienenstock, E., Doursat, R. (1992). Neural networks and the bias/variance dilemma. *Neural Comput.* 4, 1–58.

[23] Draper, N. R., Smith, H. (1998). *Applied Regression Analysis.* Hoboken, NJ: Wiley.

[24] Mazurek, P., Wagner, J., Morawski, R. Z. (2015). "Acquisition and preprocessing of data from IR depth sensors to be applied for patients monitoring," in *Proceedings of the 8th IEEE International Conference on Intelligent Data Acquisition and Advanced Computing Systems, Technology and Applications,* 705–710, Warsaw University of Technology, Warsaw.

[25] Hinton, G., and Salakhutdinov, R. (2006). Reducing the dimensionality of data with neural networks. *Science* 313, 504–507.

[26] Bengio, Y. (2009). Learning deep architectures for AI. *Foundat. and Trends Mach. Learn.* 2, 1–127.

[27] Erhan, D., Bengio, Y., Courville, A., Manzagol, P. A., Vincent, P., Bengio, S. (2010). Why does unsupervised pre-training help deep learning? *J. Mach. Learn. Res.* 11, 625–660.

[28] Golovko, V., Kroshchanka, A., Rubanau, U., Jankowski, S. (2014). "Learning Technique for Deep Belief Neural Networks," in *Proceedings of 8th International Conference on Neural Networks and Artificial Intelligence ICNNAI 2014,* Brest, Belarus. Berlin: Springer Communications in Computer and Information Sciences 440, 136–146.

[29] Scholz, M. (2012) Validation of nonlinear PCA. *Neural Process. Lett.* 36.1, 21–30.

[30] Kramer, M. A. (1991). Nonlinear principal component analysis using autoassociative neural networks. *AIChE J.* 37, 233–243.

[31] Jankowski, S., Dusza, J. J., Wierzbowski, M. Oręziak, A. (2005). "SVM Detection of premature ectopic excitations based on modified PCA", in *6th International Symposium on Biomedical Data Analysis ISBMDA 2005,* eds. J. L. Oliveiro, V. Maojo, F. Martin-Sanchez, A. Sousa Pereira, (Berlin: Springer-Verlag), Aveiro, Portugalia, Lecture Notes in Bioinformatics LNBI 3745, 173–183.

[32] Oja, E. (1989). Neural networks, principal components, and subspaces. *Int. J. Neural Syst.* 1, 61–68.

[33] Hornik, K., and Stinchecombe, M., H. (1989). White, multilayer feedforward networks are universal approximators. *Neural Netw.* 2, 359–366.

[34] Bishop, C. M. (1996). *Neural Networks for Pattern Recognition.* Oxford: Clarendon Press.

9

Decision Trees Implementation in Monitoring of Elderly Persons Based on the Depth Sensors Data

**Piotr Bilski, Paweł Mazurek, Jakub Wagner
and Wiesław Winiecki**

Institute of Radioelectronics and Multimedia Technologies,
Faculty of Electronics and Information Technology, Warsaw University
of Technology, ul. Nowowiejska 15/19, 00-665 Warszawa, Poland

Abstract

The chapter presents application of the Decision Tree (DT) to the fall detection of elderly people monitored by the infrared depth sensors. The decision making system works on data acquired by sensors, recording movement of the person, and raising the alarm if his/her behaviour suggests that the accident took place. From the measurement, various data morphological features are extracted, further processed by the DT. The classifier was adjusted to the presented problem, including the weighting of the distinguished categories. Various configurations of the DT were implemented, showing its usefulness not only to solve the presented task, but also revealing some disadvantages, compared to other methods.

Keywords: Artificial intelligence, Fall detection, Decision trees, Depth sensors.

9.1 Introduction

The importance of the computer-based monitoring systems for disabled or elderly persons is growing. The developed societies consider the help for such people their prime concern. Also, technological advancement allows for the implementation of sophisticated, automated modules working on-line, and cooperating with remote control nodes, playing role of data warehouses. The monitoring systems are rapidly developed thanks to the implementation of new measurement and data acquisition (DAQ) techniques. Based on them, autonomous modules make decision about the state of the monitored person after processing the incoming information. The key features of such solutions [1] are speed, accuracy, and ensuring the privacy of monitored people. The latter makes application of standard cameras [2], recording images in the visible spectrum, an impossible choice. Currently, the non-intrusive approaches are preferred. This includes installation of infra-red (IR) sensors [3] or radars. Both can be mounted in the direct vicinity of persons, working discretely and not disturbing them in their daily activities. The cost of DAQ devices is low enough to make them an attractive part of the complex system, introducing artificial intelligence (AI) methods, operating in the embedded hardware, such as microcontrollers.

The chapter presents the novel method for online monitoring of the elderly persons and detecting their situation (fall or normal activity) based on the information collected by the IR depth sensors. Such a solution allows for the constant observation of, for instance, lodgers in the nursing homes. Multiple problems have to be solved to create the operating framework. The first one was extracting characteristic features from the measured signals bearing the maximum amount of information about the person's actual state. The second was the selection of the appropriate decision-making module, processing features, and determining whether the person requires assistance of the medical staff, or not. From the AI point of view, this is the binary classification task, where only two states have to be distinguished.

The original input of the presented work includes introduction of the specific set of features extracted from the depth sensors and adjustment of the classifier to minimise the chance of missing the fall event (which could lead to the death of the monitored person).

The structure of the chapter is as follows. Section 9.2 contains the state-of-the art regarding the fall detection and monitoring of elderly persons. In Section 9.3 the architecture of the proposed monitoring system is presented. Section 9.4 provides the information about the measurement data acquisition and processing. In Section 9.5 features extraction from the measurement data is described. Section 9.6 introduces the decision tree (DT) classifier and its adjustment for the defined task. In Section 9.7, experimental results are provided. Section 9.8 contains conclusions and future prospects for the proposed system and required modifications to make it implementable in the actual module working with people.

9.2 Related Works

The problem of the fall detection was intensely studied in the literature. The topics covered can be decomposed into three groups: DAQ system, providing the information for the classifier, method for the feature extraction, and the identification algorithm implementation.

The DAQ module usually exploits the video stream [4] or the IR sensors [3]. This allows for creating 3D images of the scene, from which features are extracted. In the first case, the camera allows for the processing of the RGB components, which facilitates determining the position and dynamics of the human silhouette. The multiplied IR sensors provide the identification system with the similar information, but coming from another range of frequencies. The latter hide details of monitored peoples' personalities, therefore are eagerly implemented. The number of sensors is important. In most cases the standard and inexpensive equipment (MS Kinect) is used [5]. Sometimes Passive IR (PIR) arrays are exploited to increase resolution of the system [6]. The presented hardware is often supported by the auxiliary sensors, detecting pressure, or vibration (accelerometers), which introduce new features originating from different phenomena [7]. All considered acquisition methods provide enough information to further identify the fall event. Selection of IR, pressure, or vibration sensors allows for maintaining the anonymity of monitored persons.

The considered features are diverse, depending on the analysed domain or implemented signal processing operations. If the cameras or IR sensors are used, the skeletal features (such as joint locations), head position or the shape of the silhouette are monitored [8]. Their change in three dimensions over time allows for determining whether the fall occurred or not. In the IR-originated data, the Scale Invariant Feature Transform (SIFT) is used, facilitating the feature determination in the three-dimensional space. Additionally, the statistical features of image or its frequency components may be used [9]. The main problem here is the number of used features and their significance for the identification procedure. To maximise the latter, the preprocessing stage may be used, including filtering and segmentation, Spatial–Temporal Descriptors or Principal Component Analysis [10].

Identification modules are mainly based on the AI methodology. In most cases, various versions of Artificial Neural Networks (ANN) are used. Because of the high flexibility and the ability to work in the uncertainty conditions, Support Vector Machines (SVM) [11] are the primary choice, usually used on data from cameras (such as skeletal or statistical features). More traditional approaches, including the Multi-Layered Perceptron (MLP) are also applied [4]. Other methods include rule-based inference algorithms, such as DTs [12] or Fuzzy Logic (FL) [7]. Their advantage over ANN lies in the form of stored knowledge, which is here understandable by the human operator. Statistical analysis includes Naïve Bayes Classifier (NBC [12]) and Hidden Markov Models (HMM [6]). Most of presented approaches work well with the available data, giving the accuracy between 70 and 90%.

The analysis of state of the art shows additional methods should be implemented, especially in the field of features selection and identification accuracy. The latter is usually calculated as the number of correctly identified states (whether they are falls or cases of normal behaviour), while the problem here is to minimise the chance of mistaking the actual fall with anything else. This will be discussed further in detail.

9.3 Architecture of the Monitoring System

The structure of the proposed monitoring system is presented in Figure 9.1. The IR sensors are located in the vicinity of the patient, constantly gathering information about her/his condition. Their location allows for creating the three-dimensional map of the person's movement, its speed and acceleration. When the event of the possible fall is detected (usually related to the abrupt increase in the position or velocity), the decision-making module in the computer system processes the data from sensors to decide whether the event that occurred is "the fall" (labelled as "1") or not (labelled as "0"). In the first case, the alarm is raised and the medical personnel comes to help. Otherwise, the event is ignored and treated as the typical behaviour of the patient (like sitting).

To correctly execute the reasoning process during the decision making, the AI module exploits knowledge about "the fall" scenarios and all other events (further called "non-falls"). The most popular method for collecting such information is to record multiple situations for both types of events by

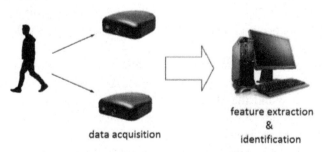

data acquisition

feature extraction
&
identification

Figure 9.1 Structure of the fall detection system.

measuring signals, from which relevant features are extracted. Afterwards, the machine learning algorithm is used to generate knowledge required for distinguishing between events belonging to both categories. The AI-based procedure is presented in Figure 9.2. The training stage consists in preprocessing of collected data (to extract relevant features) and extracting knowledge to establish dependencies between features and the category of the scenario (further called the example). Knowledge is next used to assign the actual person's condition to one of two categories. The system must be generic enough to react correctly on various events (as each person behaves differently). Because databases with the desired data are rarely available, recordings are usually performed with the help of actors or dancers, who play the role of elderly patients in the laboratory environment. This way the training L and testing T data sets are created, containing n examples e (i.e., vectors of m characteristic features s_{ij}). They are supplemented with the actual category c_i ("0" or "1"), which makes them usable in the supervised learning scheme. Both L and T have identical structure and their cardinality n depends on the applied validation procedure (see Section 9.5).

$$L = T = \begin{bmatrix} s_{11} & \cdots & s_{1m} & c_1 \\ \vdots & \ddots & \vdots & \vdots \\ s_{n1} & \cdots & s_{nm} & c_n \end{bmatrix}. \tag{9.1}$$

Because it is difficult to obtain the actual data collected by monitoring elderly people, their role is usually played by actors. The question is how their behaviour corresponds to actions of people being the target of the system. Bearing this in mind, two PhD students were subjected to recordings [13].

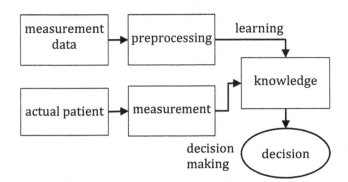

Figure 9.2 Scheme of the AI-based fall detection system.

The correctness of the decision about the patient's state must be evaluated. We use the sample error e_s, i.e., the number of the incorrectly classified examples from the testing set T.

$$e_s = \frac{|e : d(e) \neq c(e)|}{|T|} = \frac{FP + FN}{|T|} \tag{9.2}$$

$$acc = 1 - e_s = \frac{|e : d(e) = c(e)|}{|T|} = \frac{TP + TN}{|T|} \tag{9.3}$$

Alternatively, accuracy (Equation (9.3)) is the relative number of correctly classified examples from the set T.

Usefulness of this approach is limited, as two different types of possible errors must be considered. The first one is the false alarm (so-called "False Positive" – FP), i.e. the situation when the fall (event "1") is detected, while it actually did not happen. The consequence of this error is the unnecessary dispatch of the medical assistance to the patient. The second mistake is missing the actual fall (i.e., classifying the event "1" as "0", so-called "False Negative" – FN). This situation is more serious, as such error may lead to the death of the monitored person. The correct decisions belong to the "True Positive" (correct fall detection) and "True Negative" (correct detection of normal behaviour) class. During the evaluation of the intelligent method used for the task, the TP and TN ratios should be maximised, the FP ratio should be minimised and the FN ratio must be suppressed to zero at the same time. The primary aim of the implemented system is then to detect all falls, even if some false alarms will be raised.

9.4 Characteristics of the Acquired Data

This section contains the information about the procedure of collecting measurement data and preparing it for training by the DT.

9.4.1 Data Acquisition Technique

The data for testing fall detection algorithms have been acquired using two synchronised IR depth sensors, further called S_1 and S_2, respectively. The measurement devices are two Microsoft Kinect modules (model 1473). Their configuration, relative to the observed area, is presented in Figure 9.3. The distance between them was set to about 3 m. The monitored person was moving at the distance of about 1.5–5 m from each device. Subsequent

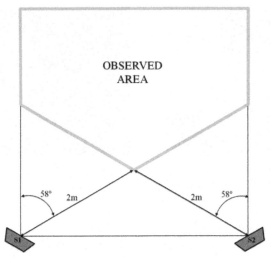

Figure 9.3 Configuration of two depth sensors (S_1 and S_2) relative to the observed area.

experiments lasted for 10 s and consisted in recording data from both sensors simultaneously, with the frame frequency of 30 fps. They resulted in a sequence of 300 depth images, i.e., 480 × 640 matrices of integer numbers representing distance from the sensors.

A set L of 18 falls and 18 non-fall scenarios has been designed. This way the set of total 144 sequences of depth images was created (each of 36 scenarios was repeated by two actors, and recorded by two sensors: S_1 and S_2).

9.4.2 Data Preprocessing

Every pixel of an image, mapped by sensors is represented by a triplet of integer numbers (i, k, d), where $i \in \{1, 2, \ldots, I\}$ is the column index, $k \in \{1, 2, \ldots, K\}$ is the row number, and $d \in \{1, 2, \ldots, 5000\}$ is the distance to the sensor (in mm). Since such a relative representation of the image may be a source of ambiguity – the same person may appear as larger or smaller, depending on the distance from the device – it was transformed to the absolute representation based on the global space coordinates (x, y, z), similarly to SIFT. The corresponding mathematical procedure consists of two operations [13]:

- identification of a set P of pairs (i, k), being the representative of the silhouette of a monitored person (Figure 9.4);
- transformation of the triplets (i, k, d), corresponding to P, into the triplets (x, y, z).

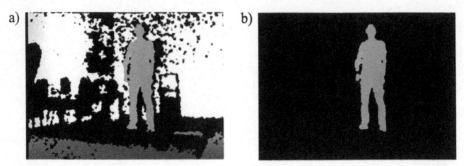

Figure 9.4   Extraction of the silhouette of a monitored person: (a) the original image recorded by the sensor S_1; and (b) the result of extraction.

The absolute coordinates $(x_{i,k}, y_{i,k}, z_{i,k})$, calculated for all $(i, k) \in P$, were used for computation of the coordinates of the silhouette "mass centre":

$$(x_C, y_C, z_C) = \frac{1}{|P|} \sum_{(i,k)\in P} (x_{i,k}, y_{i,k}, z_{i,k}), \tag{9.4}$$

and its magnitude, i.e., the effective reflection area:

$$M = \frac{1}{|P|} \left(\sum_{(i,k)\in P} d_{i,k} \right)^2. \tag{9.5}$$

The application of this procedure to a sequence of depth images, acquired at the predefined time instants $t_n(n = 1, 2, \ldots)$, resulted in four-dimensional trajectories:

$$\{x_C(t_n), \ y_C(t_n), \ z_C(t_n), \ M(t_n)\}, \tag{9.6}$$

which may be used for classification of events recorded by S_1 or S_2.

9.5 Extraction of Features

From the recorded trajectories (Equation (9.3)), six patterns were analysed for the single scenario. As opposed to other researchers [14], in this project the characteristic points (further called "morphological") were extracted from each pattern (Figure 9.5). They include:

- the value of the pattern (position or velocity in every of three dimensions) at the beginning of the event (in the 20th frame – the black point A).

- the value of the pattern at the end of the event (in the 50th frame from the end – the black point *B*).
- the maximum and minimum value of the pattern and the number of corresponding frames (points *C* and *D*, respectively).
- coordinates of two the greatest differences between the neighbouring minimum and maximum values (green points *E* in Figure 9.5).

Example of the recorded signals (position and velocity) in three dimensions during the fall event is presented in Figure 9.6. From each signal features are extracted as in Figure 9.5.

The features were extracted from the original patterns without any prior preprocessing, such as the elimination of the background noise. The applied approach was sufficient to ensure the acceptable accuracy. Alternatively, the de-noising procedures (including the polynomial and spline approximation) were considered. As they do not improve the identification accuracy, their application was eventually rejected.

The overall number of features from each of six patterns was 14 (Figure 9.5), leading to 84 features for the particular example. This way 144 training examples were created for the classifier to process. Note that the feature extraction in the actual case requires the prior event detection (i.e., suspicious behaviour of the patient, which might be his/her fall) based on the on-line analysis of the IR signals. This stage is omitted in the presented research, but must be required in the practical implementation of the system.

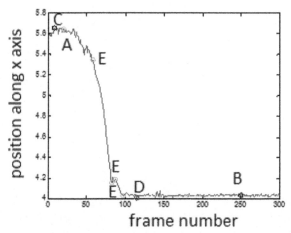

Figure 9.5 Extraction of the features from the position pattern.

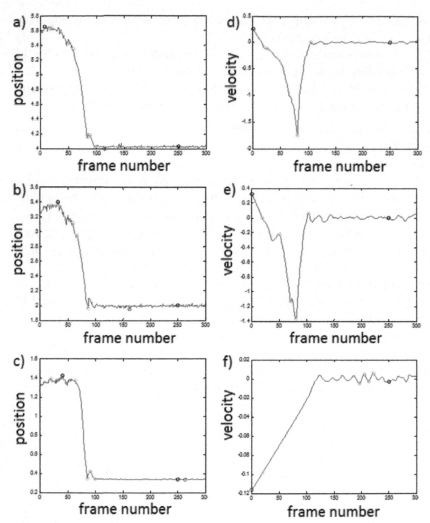

Figure 9.6   Recorded signals along axes *x*, *y* and *z* of position (*a*, *b*, and *c*, respectively), and velocity (*d*, *e* and *f*, respectively) during the fall event.

9.6 Decision Trees

This section presents the structure and implementation of the DT used in the fall detection module. As the significance of distinguished categories is not the same (missing falls should be avoided if possible), modification of the DT structure to prefer the category "1" is presented.

9.6.1 Tree Structure and Construction Algorithm

The DT [15] (Figure 9.7) is the hierarchical structure of nodes, starting from the initial root at the first level and ending with leaves, containing the particular category. Each parent node is connected to two child nodes (or leaves) belonging to the lower level. The edges connecting nodes are related to the test stored in the parent node. It is the threshold value θ of the selected feature s_i, to which the corresponding value of the example is compared. The sequence of tests enables DT to make the decision about the event category. The example "travels" from the root to one of the leaves. In each intermediate node the test is executed for the processed example. Comparison between the threshold θ and the value of the corresponding attribute s_i in the example causes its relocation to one of two child nodes. The process is repeated until the example reaches the leaf, which points at the particular category.

Application of DT requires dividing the available data D into the training and testing subsets: L and T, respectively (Figure 9.2). The Leave-One-Out (LOO) and Repeated Random Subsampling (RRSS) Cross-Validation (CV) were used to check the generalisation abilities of the classifier. In the former, D is divided into k subsets (each of the size of D/k). The operation consists in selecting one subset as T, while remaining ones become L. The assignment operation is repeated k times, resulting in k classification errors. The overall performance of DT is measured as the mean value of e_s for k trials:

$$\bar{e}_s = \frac{\sum\limits_{i=1}^{k} e_s(i)}{k}.$$ (9.7)

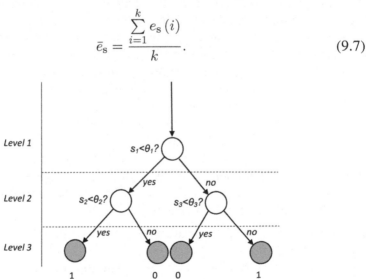

Figure 9.7 Structure of the decision tree in the binary classification problem.

Because the division of D into L and T may be conducted in multiple ways, RSSSCV should be repeated a couple of times to verify, if the random configurations of L and T content influence the value of e_s and the threat of missing the fall event. Therefore, the mean value of e_s with the standard deviations may be used to supplement the information about the accuracy (Equation (9.6)) with its repeatability.

In LOOCV, D is systematically divided into one-example testing subsets, while all remaining data belong to the L. It is the most exhaustive verification ($k = |D|$), evaluating the influence of all possible training and testing data combinations on the value of e_s.

The tree induction [16] is applied to extract knowledge from the measurement data. In each step of this recurrent process, the node with the threshold θ, or the leaf is created. The node divides the currently processed set L into two subsets of similar cardinality in such a way that each category is present only in one of them. This way examples are sorted according to their categories using the minimal number of nodes. If the current subset contains only examples belonging to one category, the further division is not needed. Instead, the leaf is created, representing this category. The training process is susceptible to overlearning. All examples from the training set may be separated correctly, but the performance on the testing set will be low. This means that the DT has limited (evaluated with CV) generalisation abilities.

The entropy was used to select the best candidates for the threshold. This is the measure of disorder in data, with high value for the threshold θ creating two subsets, which contain the same categories, distributed evenly (in this case, both contain examples labelled with "0" and "1"). The low entropy value is for the thresholds dividing examples into two subsets with mutually excluding categories. This is preferred, as it partitions data into separate categories in the shortest amount of time, leading to the simplest tree structure. Candidates for thresholds (labeled as t_{ij}) are calculated as the middle points between neighbouring values of sorted symptom values (labeled as o_{ij}). Next, one of them with the minimum entropy is selected. Because there are multiple candidates for the threshold in each node with equal, minimum entropy value, various strategies of the selection may be applied. In the presented research, the following methods were implemented:

(a) selection as the threshold the attribute with the largest distance from the neighbouring values;
(b) selection as the threshold the attribute with the smallest distance from the neighbouring values;

(c) selection as the threshold the attribute occurring in the tree nodes the greatest number of times;
(d) selection as the threshold the features occurring in the tree nodes the smallest number of times; and
(e) random selection of the feature from the subset of the ones with the smallest entropy.

Influence of the selected strategy on the obtained classification accuracy is discussed in the "Experimental Results" section.

9.6.2 Tree Modification to Maximise Accuracy

The presented classification problem requires that both categories are treated differently, as missing the actual fall has more serious consequences than raising the false alarm. In most approaches (except, for instance, k-Nearest Neighbours [17]), weighting categories' importance is difficult. In the DT case, it can be done by modifying the tree structure after it was created. The candidates t_{ij} for the test value in each node are located in the middle of their two neighbours, being the ordered values ($o_{i,j}$ and $o_{i,j+1}$) of the particular feature s_i:

$$t_{ij} = \frac{o_{i,j} + o_{i,j+1}}{2} \qquad (9.8)$$

In the penultimate level of the DT this relation can be modified to favour one of the categories. Instead of creating the tested attribute's threshold value in the middle between two neighbours, it can be moved towards one of them, according to the value of the coefficient $\alpha \in (0, 1)$:

$$\theta_{ij} = \frac{o_{i,j} + o_{i,j+1}}{o_{i,j} + \alpha \left(o_{i,j+1} - o_{i,j}\right)}. \qquad (9.9)$$

If the i-th attribute's j-th value represents the example belonging to category "0" and its j + 1st value is for the example belonging to category "1", then the latter can be supported by moving the threshold closer to the value $o_{i,j}$ (by setting $\alpha < 0.5$). This increases the chance of selecting the category "1" in the border area between falls and non-fall events. Similarly, if the j-th ordered value belongs to the example from category "1", the threshold should be moved to the example $o_{i,j+1}$ (by setting $\alpha > 0.5$). The idea is presented in Figure 9.8, where areas of the particular response of the classification system (h) are indicated, depending on the selected threshold θ_{ij} position. Moving the latter towards the non-fall event increases the range of feature's values leading to the fall detection. The key problem is determination of the α value.

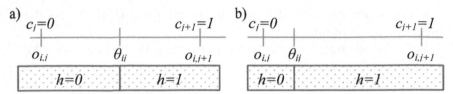

Figure 9.8   Illustration of the influence of the threshold position on the ability to detect the fall event: equal weighting (a) and modified, preferring the fall (b).

In the presented research the best results were obtained for $\alpha = 0.3$ or $\alpha = 0.7$, depending whether $o_{i,j}$ or $o_{i,j+1}$ represents the example belonging to the category "0".

9.7 Experimental Results

The conducted experiments included verification of the decision tree parameters and its ability to suppress the FN ratio (missing actual falls). Comparison between the classification outcomes for various strategies is in Table 9.1, where "$1 - e_s$" is the mean percentage of the correct classifications. Columns "FP" and "FN" present the percentages of false alarms and missed falls, respectively. Three CV methods (RRSSCV for $k = 5$ and 10 and LOOCV) were considered. Criteria (a) and (d) are the most promising, ensuring the greatest separation in the training data and providing the greatest variability in features. They not only provide the smallest sample error, but also ensure the minimal number of falls missed. The percentage of false alarms is also small, although this error is of secondary importance. The detailed results for the threshold selection strategy (a) are in Table 9.2, while for the criterion (c) – in Table 9.3. They show the contrast in accuracy between the best and the worst option and their relation with the tree structure.

The comparison between the CV approaches shows the LOOCV method is the most reliable, as it does not depend on the random process of selecting

Table 9.1   Comparative analysis of the fall detection accuracy regarding various node construction strategies (LOOCV)

Strategy	$1 - e_s$ (%)	FP (%)	FN (%)
(a)	93,75	4,85	1,13
(b)	87,58	7,28	4,62
(c)	77,13	10,46	12,39
(d)	89,33	8,16	2,13
(e)	88,19	9,02	2,77

Table 9.2 Cross-validation results (mean values for RSSSCV) for the decision tree with thresholds selected using the maximum distance from separated attribute values (criterion (a))

CV Type	$1 - e_s$ (%)	FP (%)	FN (%)
LOOV	93,75	4,85	1,13
RSSCV-5	92,58	5,61	1,81
RSSCV-10	89,33	7,3	3,3

Table 9.3 Cross-validation results (mean values for RSSSCV) for the decision tree with thresholds selected using the maximum number of occurrences of this threshold in the tree (criterion (c))

	$1 - e_s$ (%)	FP (%)	FN (%)
LOOV	77,13	10,46	12,39
RSSCV-5	85,55	10,0,4	4,11
RSSCV-10	87,33	6	6,66

examples to training and testing subsets. Although the most time-consuming, it also shows the performance of the classifier working on the largest available set *L*. On the other hand, variants of the RRSS verification depend on partitioning of examples into *L* and *T* and will be different each time. To suppress randomness between trials, every RRSSCV experiment was repeated five times, resulting not only in the mean classification accuracy, but also standard deviation, representing variability in particular trials. Mean classification accuracies for the single experiment regarding the strategy (c) are in Figure 9.9, while standard deviations of these results are in Figure 9.10.

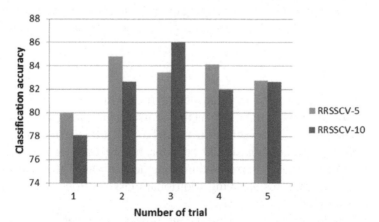

Figure 9.9 Classification outcomes for subsequent trials in two RRSSCV schemes (strategy (c)).

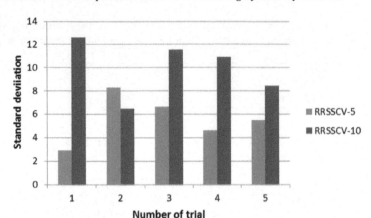

Figure 9.10　Standard deviations of classification outcomes for subsequent trials in two RRSSCV schemes (strategy (c)).

The standard deviation is relatively high, proving that LOOCV is the most reliable validation approach. Also, there are no significant differences between the different values of k. The greatest impact on the obtained accuracy has the selected node generation strategy. The problem of separating between two categories of different weights (as missing the fall has more serious consequences than the false alarm) remains, but it now is minimal and has smaller impact on the overall fall detection accuracy, than for the original DT algorithm. The price for the improved resolution of the fall detection is the increased number of false alarms. The generated DT have a complex structure for the binary classification task, containing even 47 nodes (including leaves). This may sometimes lead to the overlearning, minimised by the tree pruning.

9.8 Conclusion

The decision tree is the fast and accurate method for the fall detection based on the IR sensors measurement data. To maximise its performance, the method of the threshold for the node selection must be carefully designed. Also, the training data preprocessing should be done before training the tree to decrease the number of false alarms and missing actual falls. The considered task is complex and requires applying additional classification methods and considering various sets of features.

As the selected features have a significant impact on the behaviour of the decision making module, various attributes should be checked and compared (including, for example, spectral or cepstral information). The second problem to solve regarding the DT application was the introduction of the category weights, allowing for detection of all actual falls (i.e., FN = 0). The tree itself is unable to identify the varying importance of categories; therefore, separating areas between the categories were modified (by moving thresholds towards the examples labelled with the category "0"). This decreases the number of missing fault events, but sometimes increases the number of false alarms.

The future research should also include the comparative analysis between the DT and other classifiers, including random forest, artificial neural network or k Nearest Neighbours. The classifier fusion (including various methods working in parallel) must be considered as well.

Acknowledgment

This work has been accomplished within the project PBS2/A4/0/2013 "The Non-invasive System for Monitoring and Analysis of Electricity Consumption in the Area of the End-user" financially supported by the National Centre for Research and Development.

References

[1] Zhou, Z., Stone, E. E., Skubic, M., Keller, J., and He, Z. (2011). "Nighttime in-home action monitoring for eldercare," in *33rd Annual International Conference of the IEEE EMBS*, 5299–5302, Boston, Massachusetts.

[2] Kepski, M., Kwolek, B. (2014). "Fall Detection Using Ceiling-Mounted 3D Depth Camera," in *International Conference on Computer Vision Theory and Applications VISAPP 2014* (Setúbal: SciTePress), Vol. 2, 640–647.

[3] Mastorakis, G., and Makris, D. (2012). Fall detection system using Kinect's infrared sensor. *J. Real Time Image Process.* doi: 10.1007/s11554-012-0246-9.

[4] Foroughi, H., Shakeri Aski, B., and Pourreza, H. (2008). "Intelligent Video Surveillance for Monitoring Fall Detection of Elderly in Home Environments," in Proceedings 11th International Conference on Computer and Information Technology (ICCIT 2008), Khulna, Bangladesh, pp. 219–224.

[5] Gasparrini, S., Cippitelli, E., Spinsante, S., and Gambi, E. (2014). A depth-based fall detection system using a kinect sensor. *Sensors*, 14, 2756–2775. doi: 10.3390/s140202756

[6] Popescu, M., Hotrabhavananda, B., Moore, M., and Skubic, M. (2012). "VAMPIR: An automatic fall detection system using a vertical PIR sensor array," in Proceedings of 6th International Conference on Pervasive Computing Technologies for Healthcare (PervasiveHealth) and Workshops, 163–166.

[7] Kepski, M., Kwolek, B., and Austvoll, I. (2012). *"Fuzzy Inference-Based Reliable Fall Detection Using Kinect and Accelerometer,"* in ICAISC 2012, Part I, LNCS 7267, 266–273.

[8] S. Sinha, S. Deb, Depth sensor based skeletal tracking evaluation for fall detection systems. Int. *J. Comput. Trends Technol.* 9, 350–354.

[9] Nghiem, A. T., Auvinet, E., Meunier, J. (2012). "Head detection using KINECT camera and its application to fall detection," in *Proceedings of 11th International Conference on Information Sciences, Signal Processing and Their Applications*, 164–169.

[10] Hazelhoff, L., Han, J., and de With, P. H. N. (2008). *Video-based fall detection in the home using principal component analysis.* Lecture Notes in Computer Science, Vol. 5259, 298–309.

[11] Charfi, I., Miteran, J., Dubois, J., Atri, M., and Tourki, R. (2013) Optimized spatio-temporal descriptors for real-time fall detection: comparison of support vector machine and Adaboost-based classification. *J Electron. Imag.* 22, 041106. doi: 10.1117/1.JEI.22.4.041106.

[12] Delahoz, Y. S., and Labrador, M. A. (2014). Survey on fall detection and fall prevention using wearable and external sensors. *Sensors*, 14, 19806–19842, doi: 10.3390/s141019806.

[13] Mazurek, P., Wagner, J., and Morawski, R. Z. (2015). Acquisition and preprocessing of data from IR depth sensors, developed for a system of patients monitoring, in *8th IEEE International Conference on Intelligent Data Acquisition and Advanced Computing Systems: Technology and Applications*, Warsaw, Poland, 705–710.

[14] Davari, A., Aydin, T., and Erdem, T. (2013). "Automatic fall detection for elderly by using features extracted from skeletal data," in *International Conference on Electronics, Computer and Computation (ICECCO)*, 127–130, Ankara, Turkey.

[15] Nguyen, N.-T., and Lee, H.-H. (2008). Decision tree with optimal feature selection for bearing fault detection. *J. Power Electron.* 8, 101–107.

[16] Jamehbozorg, A., and Shahrtash, S. M. (2010). A decision-tree-based method for fault classification in single-circuit transmission lines. *IEEE Trans. Power Deliv.* 25, 2190–2196.

[17] Bilski, P. Mazurek, P., and Wagner, J. (2015). "Application of k Nearest Neighbors Approach to the Fall Detection of Elderly People Using Depth-Based Sensors," in *Proceedings IDAACS 2015 Conference*, 24–26 September, Warsaw, Poland, 733–739.

10

Recurrent Approximation in the Tasks of the Neural Network Synthesis for the Control of Process of Phototherapy

Alexander Trunov

Medical Equipment and System Dept., Petro Mohyla Black Sea National University, Mykolayiv, Ukraine

Abstract

The chapter reveals the problem of electromagnetic wave (EMW) interaction in the case when the electron of radical transition. The task of interaction between photon and electron with three natural frequencies different in three directions for terminal enzyme of respiratory chain cytochrome-c-oxidase or other photo acceptors was considered and solved. The solution of problems of encoding and decoding of polarized light interferogram data and calculation of radiation dose for different spectral composition are proposed. The examples of problem of analytical learning as standard elements for design multilayers artificial neural network (ANN) with any complicated configuration of input–output vector-function are considered. The vector-indicator was introduced and a new approximation–expansion within these three levels digital quantities was proposed. Recurrent approximation was applied for synthesis of recurrent artificial neural network (RANN). Solutions for synaptic weight coefficients (SWC) as recurrent sequence are represented in analytic form. Synthesis of control system as mini–max problem for efficiency optimization was solved. Convergence of recurrent approximation for two schemes of approach by a linear and quadratic curve are proposed and discussed for System Support Decision Making (SSDM) of physiotherapy with ANN and RANN.

Keywords: Phototherapy, EMW interaction, Concentration distribution, Irradiation dose, Neural network, Analytical learning, Solutions – recurrent sequence, Convergence, SSDM.

10.1 Introduction

The modern state of knowledge about interaction betweeen electromagnetic wave (EMW) and electron of substance [1–8, 10–14, 24–31] and models which was different due to differences in its structure [4–6] and the nature [8] was not exhaustive and universal. Particularly it was discovered during investigation of nonliving and wildlife matter, normal cells and abnormal cells of tumor [1–3, 23–31]. Simultaneously, reconsideration of theoretical success in the field of comparative analysis for dynamic system by using decision-tree induction [19, 35] and revision of the present progress in low intensive light therapy, including technological success in development of light source and devices for hyper spectral analysis [37, 43], result applications in magnetic resonance imaging [14, 23–26, 38, 39] at calibration and parametrization [34] of sensor [38, 39, 43] are stimulated reformulation of approaches to early diagnosis, prevention and treatment [9–14].

The latest results in the field of development means of automation [14–19, 24, 27–29, 31–33], and design of structural components [10–14, 39], and units for adaptive control [18, 23, 31] with regulators based on an artificial neuron nets (ANN) [14–18, 22, 23, 42] was a reason for a new implementation of this modern experience as instruments of creating data base and base of knowledge. However, sexes of realization modern achievements for control of phototherapy cannot be an implement in medical practice without new technologies for teaching personals [40] and also without means of increasing the resistance of computer systems for towards of virus attacks [36]. All of this aspect activate search of new paradigms of physiotherapy [11–13]. Success of these attempts can be reached only due to mutual investigation of peculiarities of the EMW with bio-tissue and tool for early diagnosis, prevention and treatment and development of modern means for control based on the latest results in perfection of ANN [14–19, 21–23].

At present, it is well known and proved and well documented that the action of coherent with low level of intensity and not coherent radiation with the same spectral composition, intensity and dose makes the same biological response for various bio objects [1–3]. In the case of low intensive laser therapy [4–6], additionally to the main acknowledged mechanism of therapeutic action – the absorption by chromophores can be observed in deep layers as a result of random interference of rays and as mechanism of cellular absorption. As it was shown in References [1–3] and in some of its reference are indicated the two type of photo acceptors and enzymes. The first type was cytochrome-c-oxidase for visible and near infrared ranges. The second was dehydrogenase enzymes, regarded as other type of photo acceptor, but only for blue-violet range of the spectrum [3].

Other mechanisms are discussed (e.g. [2, 3]) and considering the role of enzymes into the respiratory chain and in cascade of reactions are transferred from subsequent amplification of signals into cells. Based on the mechanisms of reduction-oxidation and regulation of metabolic activity can be explained by the main contradiction of cellular effects for low intensive laser or light phototherapy. However, despite the existing wide range of publications and their reviews [3], as well as some attempts [6, 7, 9–13] to number of which can be related to a mathematical model, that can be described in the specific features of the mechanism of photon interaction such as the impact orientation of magnetic field and the orientation of polarization plane as the primary features for activating the photon absorption and the secondary infrared radiation [31]. Especially, an actual model was in the design of devices surveillance, early diagnosis and treatment [10–12, 19]. In addition, it was necessary to notice that experimantal and theoretical reseach of biological structures which contains the elements with properties of diodes or transistors and other and others schemes which assembled from them leads us to similar tasks [9]. Thus, the main unresolved problem is the lack of knowledge about specific features of fundamental properties that uniquely characterize the structure and condition of biological tissue. This specific particularity of phenomena complicates the creation of mathematical models and makes impossible to the determine the scientifically based requirements for additional effects into biological tissue with structural changes. The goal of this research is to develop by the methods of classical electrodynamics and build a mathematical model, which takes into account the impact of additional influences on the characteristics of structural transformations of cells into biological tissue under red and infrared irradiation during cycle of phototherapy.

10.2 Pointing the Task of Interaction between an Electron of Radical and Photon into Magnatic Field

Let us consider the molecular structure with carbon–hydrogen, a long chain and with one or more radicals. Suppose that the electron of radical for terminal enzyme of respiratory chain cytochrome-c-oxidase is moving and its position is determined by the radius vector $\bar{r} = f(x, y, z, t)$ at the moment of time, t. Noted that this electron is moving in a stationary magnetic field with intensity $\bar{H}$ under irradiation of EMW described as

$$\begin{cases} \bar{E}_e(t) = [E_{ex}, E_{ey}, E_{ez}]^T e^{i\omega t} \\ \bar{H}_e(t) = [H_{ex}, H_{ey}, H_{ez}]^T e^{i\omega t}, \end{cases} \tag{10.1}$$

where $\bar{E}_e(t)$ is a vector-function of the electric field intensity and E_{ex}, E_{ey}, E_{ez} are its component, $\bar{H}_e(t)$ is a vector-function of the magnetic field intensity and H_{ex}, H_{ey}, H_{ez} are its component, e is the basis of natural logarithm, and i is the complex unit.

Let us consider the photon, which was described as the light vector by Equation (10.1) and its effecting on the electron as EMW with variable wavelength, its range from ultraviolet to infrared and under the conditions of the adiabatic approximation, then the motion of electron is satisfying the equilibrium condition

$$m\ddot{\bar{r}} + k_f \dot{\bar{r}} + k\bar{r} = -\mu\mu_0 q_e \bar{v} \times (\bar{H} + \bar{H}_e e^{i\omega t}) - q_e \bar{E}_e e^{i\omega t}, \tag{10.2}$$

Equation (10.2) was written in vector form can be rewritten as the system of three scalar equations

$$\begin{cases} \ddot{x} + 2\beta_x \dot{x} + \omega_{0x}^2 x = -2\left(\Omega_z + \Omega_{ez} e^{i\omega t}\right)\dot{y} + 2(\Omega_y + \Omega_{ey} e^{i\omega t})\dot{z} - \\ \qquad\qquad - E_x e^{i\omega t} q_e/m \\ \ddot{y} + 2\beta_y \dot{y} + \omega_{0y}^2 y = -2\left(\Omega_x + \Omega_{ex} e^{i\omega t}\right)\dot{z} + 2\left(\Omega_z + \Omega_{ez} e^{i\omega t}\right)\dot{x} - \\ \qquad\qquad - E_y e^{i\omega t} q_e/m \\ \ddot{z} + 2\beta_z \dot{z} + \omega_{0z}^2 z = -2\left(\Omega_y + \Omega_{ey} e^{i\omega t}\right)\dot{x} + 2\left(\Omega_x + \Omega_{ex} e^{i\omega t}\right)\dot{y} - \\ \qquad\qquad - E_z e^{i\omega t} q_e/m \end{cases},$$
$$\tag{10.3}$$

where $k_{fj}, k_j, q_e, m, \mu, \mu_0$ are the coefficient of friction, stiffness, electron charge, mass, relative permeability and permeability of free space, respectively. The projection on the relevant axis of vector angular velocity of Larmorov's precession and damping coefficient and natural frequency of oscillation in different directions for these notations can be additionally determined

$$\beta_j = k_{fj}/(2m); \quad \omega_{0j} = (k_j/m)^{1/2}; \quad j = \overline{x, y, z.};$$

$$\Omega_j = GH_j; \quad \Omega_{ej} = GH_{ej}; \quad G = \mu\mu_0 q_e/2m^{-1}.$$

By comparing the values of terms in RHS of Equations (10.2) and (10.3) are considered as a relation between amplitude and EMW:

$$E_e\sqrt{\varepsilon\varepsilon_0} = H_e\sqrt{\mu\mu_0}.$$

In the result of consideration and calculation should be concluded: terms with multiplier Ω_{ej} is negligible small, since

$$H_e = nE_e\sqrt{\frac{8,85 \times 10^{-12}}{6,28 \times 10^{-7}}} = 3,75 \times 10^{-3}nE_e,$$

where n is the index of refraction.

The general solution of the system (10.3) seeking in a form

$$\bar{r} = [a, b, c]^T e^{iwt}. \tag{10.4}$$

The amplitudes are noticed *as a; b; c* and they are supposed to be the complex value. By substituting the partial solution (10.4) in the form of its RHS to system (10.3), we get

$$\begin{cases} d_x a = -2\Omega_z iwb + 2\Omega_y iwc - E_{ex}q_e/m \\ d_y b = -2\Omega_x iwc + 2\Omega_z iwa - E_{ey}q_e/m \\ d_z c = -2\Omega_y iwa + 2\Omega_x iwb - E_{ez}q_e/m \end{cases}, \tag{10.5}$$
$$d_j = \left(w_{0j}^2 - w^2 + i2\beta_j w\right); \quad j = \overline{x, y, z}.$$

The solution of system (10.5), in accordance with invention [10–12], the author which proposed to apply polarized EMW into a specially oriented static inter magnetic field, which has been based on the theory, as a result of consideration of motion of electron and interacting with them. Therefore $\Omega_x \neq 0, \Omega_y \neq 0, \Omega_z \neq 0$, so, we obtain

$$c = \frac{q_e}{m}\left[\left(\Omega_y E_{ey} + \Omega_z E_{ez}\right)\left(d_x d_y - 4\Omega_z^2 w^2\right) - \left(d_x E_{ey} + 2\Omega_z w E_{ex}i\right)\right.$$
$$\left(d_y\Omega_y - 2\Omega_x\Omega_z wi\right)\right] \times \left[2\left(2\Omega_y\Omega_z w^2 + d_x\Omega_x wi\right)\left(d_y\Omega_y - 2\Omega_x\Omega_z wi\right) - \right.$$
$$\left. - \left(d_z\Omega_z + 2\Omega_x\Omega_y wi\right)\left(d_x d_y - 4\Omega_z^2 w^2\right)\right]^{-1},$$

$$b = -\frac{2\left(2\Omega_y\Omega_z w^2 + d_x\Omega_x wi\right)c + \left(d_x E_{ey} + 2\Omega_z w E_{xe}i\right)\frac{q_e}{m}}{\left(d_x d_y - 4\Omega_z^2 w^2\right)},$$

$$a = \frac{2\Omega_y iwc - 2\Omega_z iwb - E_{ex}q_e/m}{d_x}.$$

For the condition when the direction of magnetic field was oriented perpendicularly to the plane of oscillation photons and also for the system of coordinate was oriented so that the vector $\bar{H}$ was oriented in parallel to the direction of the Z-axis, then the parameters $\Omega_x = 0$, $\Omega_y = 0$. For these case we have:

$$a = -\frac{2\Omega_z i\omega b + E_{ex}q_e}{md_x};$$

$$b = -\frac{q_e}{m}\left(d_x E_{ey} + 2\Omega_z \omega E_{ex}i\right)\left(d_x d_y - 4\Omega_z^2 \omega^2\right)^{-1}; \qquad (10.6)$$

$$c = -\frac{q_e E_{ez}}{md_z},$$

where c is the amplitude of oscillation in $\bar{H}$ direction which does not change. The amplitude c is a result of interaction electron of radical with inter magnetic field was determined only by module of projection of electric field intensity onto Z-axis. In other words, c is proportional to the value of module E_z – projection of EMW amplitude. The values of other two amplitudes (a and b) are changed in dependence with inter magnetic field, which was proportionally increased by its values. On the basis of analysis oscillation amplitude (Equation (10.6)), could be claimed that the stationary magnetic field reduces the energy of quantum, which is need for electron detachment and initiation of the process of structural transformation and beginning of hemostasis. In addition, it should be noted that the irradiation done by plane polarized wave, for which the oscillation plane is perpendicular to the vector, $\bar{H}$, also leads to the emission of plane polarized secondary wave in the plane of its reflected and refracted rays directly with maximally possible amplitude, i.e., energy. Fluctuations of frequencies are symmetrically shifted to the naturals frequencies $\omega_{0x}, \omega_{0y}, \omega_{0z}$ and scattered rays are plane polarized which is perpendicular to the direction of vector, $\bar{H}$, when $E_x = 0$, and $E_z = 0$. In direction of observation is parallel the vector, $\bar{H}$, scattered rays with frequencies are shifted to the naturals frequencies are plane polarized. Therefore, the structural changes of features into the molecular of bio-tissues cells or injection of micro doses of photosensitive material creates the reasons for changes in biological responses and deviations of absorption spectrum composition. Finally, as it is shown in References [1–3], is the result of electron excitation chromophore CuA and CuB in the molecule of terminal enzyme of respiratory chain cytochrome-c-oxidase.

However, the author did not investigated and did not apply quantitative analysis and mathematical modeling in the concept of quantitative variable implemented for descriptions of this effect. The initiation of secondary fluorescent radiation as result of low intensive laser or light bio stimulation and acting of stationary magnetic field is the major feature of phenomena for developing of instrument for diagnostic. The References [7, 8] were estimated to impact a stationary magnetic field. In their present work, they are evaluated the effects of electron interaction for simultaneous irradiation by photons in stationary magnetic field. On the basis of comparative analysis for different part of absorption spectrum composition and frequency content of fluorescent radiation and on the basis of discovered additional properties of symmetricity can be checked out by two properties: (i) existence of bands and their symmetricity is relatively to the natural frequency; (ii) symmetricity and dynamic are two-sided shifts as a result of programmed changing of a stationary magnetic field with strength of $\bar{H}$, have been forming two approaches for creating of algorithms of early diagnostics, prophylaxis and treatment [10–12].

10.3 Encoding and Decoding

The basic principle of sensors designing are realized in References [11–14, 24–28, 37–39, 43] due to transforming of plane polarized mode light fluxes into special image with distributed intensity as a result of frequency dependence from theirs spectrum composition.

The description of main idea was applied for method of analysis and detection of influences of external factors due to processing of images of interferogram. This type of images will be divided into three cases. The first type can be taken as result of light reflection of internal source from patients' body. The second can be received as infrared emission of certain point of body. Third type can be taken as result of light reflection of internal source rays from patients' body and secondary emission of certain point of this body. After transformation emission was introduced as a specific cosine transform for one- or two-direction. Under these conditions, the norm for them will be taken as the maximum of modulus between two or more patterns was determined by conversion coefficients with an addition of two symmetrical fringes with the two frequency in the space of pattern which will have an equal magnitude of difference for each frequency. The features of pattern defined for calibration transform for monochromatic EMW with known frequency:

in two-direction

$$\bar{B}(\omega_k, \bar{x}, \bar{y}) = \frac{B_\nu(\omega_k, \bar{x}, \bar{y})}{B_{max}} =$$

$$= \frac{ab}{B_{max}} \int\limits_{\bar{y}-\frac{\lambda_q}{2}}^{\bar{y}+\frac{\lambda_q}{2}} \left[\int\limits_{\bar{x}-\frac{\lambda_k}{2}}^{\bar{x}+\frac{\lambda_k}{2}} \cos(\omega_q \bar{y}) \cos(\omega_k \bar{x}) F(\bar{x}, \bar{y}, \omega) \, d\bar{x} \right] d\bar{y};$$

$$B_{max} = \left[\int\limits_0^b \int\limits_0^a F^2(\bar{x}, \bar{y}, \omega) \, d\bar{x} d\bar{y} \right]^{1/2}; \qquad (10.7)$$

in one-direction

$$\bar{B}(\omega_k, \bar{x}, \bar{y}) = \frac{B_\nu(\omega_k, \bar{x}, \bar{y})}{B_{max}} =$$

$$= \frac{a}{B_{max}} \int\limits_{\bar{x}-\frac{\lambda_k}{2}}^{\bar{x}+\frac{\lambda_k}{2}} \cos(\omega_k \bar{x}) F(\bar{x}, \bar{y}, \omega) \, d\bar{x}; \omega_k = 2\pi\nu_k:$$

$$B_{max} = \left[\int\limits_0^a F^2(\bar{x}, \bar{y}, \omega) \, d\bar{x} \right]^{1/2} \qquad (10.8)$$

If the interference patterns was distributed in the space for the frequency ν_k at the point with coordinates (x, y, z, ν_k, t) and at the moment of time t, its transformation is in the x-, y-direction, is performed like Equation (10.7), for each frequency of spectrum composition into the segment $\forall x \in [0, a]$, $\forall y \in [0, b]$, then the distribution function can be build as follows

$$f(x, y, z, \nu_k, t) = \frac{\partial^5}{h\nu \partial x \partial y \partial z \partial \nu \partial t} \left\{ \frac{B_\nu(x, y, z, \nu_k, t)}{B_{max}} \right\}. \qquad (10.9)$$

The process of search for a specific frequency, which may characterize the structure of cell bio-tissues can be done as it shown in Reference [31] by expanding the cosine transform of experimental interferogram image, which was simultaneously taken by using the recurrent approximation [28, 30]. Thus, if the determination of frequency values are done and interferograms are received by calibration for these values of wave length, then they proposed an expression for wave number and concentrations of cells with abnormal or new normal structural formations will be written by recurrent approximation

$$\bar{B}\left(\omega_k\right) + \Delta k_n \sum_{m=1}^{M} \frac{\partial^m \bar{B}\left(\omega_k\right)}{m!\partial k^m} \Delta k_{n-1}^{m-1} =$$

$$= \frac{ab}{B_{max}} \int_0^a \int_0^b \cos\left(\omega_q \bar{y}\right) \left\{ \begin{bmatrix} \cos\left(\omega_k \bar{x}_0\right) F\left(\bar{x}_0\right) + \\ + \cos\left(\omega_k \bar{x}_0\right) \Delta x_n \frac{\partial^m \bar{F}\left(\bar{x}_0\right)}{\partial x^m} - \\ - \left(\omega_{k0} \Delta x_n + \bar{x}_0 \frac{\partial \omega_{kn}}{\partial k} \Delta k\right) \\ \sin\left(\omega_k \bar{x}_0\right) F\left(\bar{x}_0\right) \end{bmatrix} \right\} d\bar{y} d\bar{x},$$

(10.10)

where $k = \frac{2\pi}{\lambda} = \frac{2\pi\nu}{c'} = \frac{\omega}{c'}$ is the wave vector, c' is the speed of light in vacuum, B_{max} is the norm of pattern, which is specially determined as described in Equation (10.7) and Frashe's derivatives. When $\bar{B}(\omega_k)$ is a function of wave vector is known from data base then the new value of wave vector is a result of structural changes into patient bio-issues in the form of recurrent sequence is written as

$$k_{n+1} = k_n + \left\{ \frac{a}{B_{max}} \int_0^a \int_0^b \cos\left(\omega_q \bar{y}\right) \begin{bmatrix} \cos\left(\omega_k \bar{x}_0\right) F\left(\bar{x}_0\right) + \\ + \cos\left(\omega_k \bar{x}_0\right) \Delta x_n \frac{\partial^m \bar{F}\left(\bar{x}_0\right)}{\partial x^m} - \\ - \omega_{k0} \Delta x_n \sin\left(\omega_k \bar{x}_0\right) F\left(\bar{x}_0\right) \end{bmatrix} \right.$$

$$d\bar{y} d\bar{x} - \bar{B}\left(\omega_k\right) \Bigg\} \times \left[\sum_{m=1}^{M} \frac{\partial^m \bar{B}(\omega_k)}{m!\partial k^m} \Delta k_{n-1}^{m-1} + \right.$$

$$\left. + \frac{ab}{B_{max}} \int_0^a \int_0^b \left\{ \cos(\omega_q \bar{y}) \left[\bar{x}_0 \frac{\partial \omega_{kn}}{\partial k} \sin(\omega_k \bar{x}_0) F(\bar{x}_0) \right] \right\} d\bar{y} d\bar{x} \right]^{-1}.$$

For two-direction transformation, the expressions for concentration C can be represent as recurrent sequence:

$$\bar{B}(\omega_k) + \Delta C_n \sum_{m=1}^{M} \frac{\partial^m \bar{B}(\omega_k)}{m!\partial C^m} \Delta C_{n-1}^{m-1} =$$

$$= \frac{ab}{B_{max}} \int_0^1 \int_0^1 \cos(\omega_q \bar{y}) \cos(\omega_k \bar{x}) F(\bar{x}, \bar{y}, w) d\bar{y} d\bar{x}; \qquad (10.11)$$

When $\bar{B}(\omega_k)$ is a function of frequency and concentration is known from data base then the new value of concentration is a result of structural changes into patient bio-issues in the form of recurrent sequence was determined:

$$
C_{n+1} = C_n + \left[\frac{ab}{B_{max}} \int_0^1 \int_0^1 \cos(\omega_q \bar{y}) \cos(\omega_k \bar{x}) F(\bar{x}, \bar{y}, \omega) d\bar{y} d\bar{x} \right] \times
$$

$$
\times \left[\sum_{m=1}^{M} \frac{\partial^m \bar{B}(\omega_k)}{m! \partial C^m} \Delta C_{n-1}^{m-1} \right]^{-1}.
$$

It denotes that this new algorithm was based on new instrument for analysis of interferograms [11–13, 31], determines the concentration ratio of abnormal cells as the recurrent sequence. In the event, that such specific frequency is exist, then the amplitude of transformation spectrum of its secondary emission is proportional to the concentration of structural formations, and in the case, the formation of a magnetic field in the picture of spectrum are symmetrically shifts bands for these frequency and shifting distance are proportional to the static magnetic field intensity [11–13, 31].

10.4 Specific Features of Dose Calculation and Formation of the Spectral Composition of Radiation

To ensure the prevention and treatment by the methods of resonant conformational therapy it is necessary to calculate the parameters of impact [10, 31]. The main quantities had to be used as components of input vectors x_{im} can be received by data processing of signals from cameras and hyper spectral sensors: magnitude of intensity, dose, irradiation area and the velocity of light bands propagation. The intensity $I(x, y)$, dose D and biological result, N_e of interaction of irradiation on a surface, S of irradiation of biological tissues with $N_\nu(x, y, t)$, photons at point (x, y, z) coordinates in moment of time, t will be determined:

$$
I(x, y) = \int_{\nu_1}^{\nu_2} f(x, y, z, \nu, t) h\nu d\nu; \quad D = \iint_{V,t} I(x, y) dV dt;
$$

$$
N_e = \int_0^t \int_0^\infty \iiint_v \rho(\nu) \, exp(-\gamma z) f(x, y, z, \nu, t) dx dy dz d\nu dt.
$$

where N_e are defined as earlier and $\rho(\nu)$ is the number and probability of electron detachment from cytochrome-c-oxidase into mitochondrion cell, $f(x, y, z, \nu, t)$ is the spectral distribution of energy density per meter is an inside of biological object, γ is the coefficient of energy absorption and scattering.

10.4.1 Application of Data Mining for Decomposition of Scalar- or Vector-Function of Vector

Let us introduce a productive rule for three levels of triggering comparators. Suppose that an infinite set of real numbers $\forall Y \in (-\infty, \infty)$ exists with a set of three standard values Y_1, Y_2, Y_3, then we introduce a comparator which was realized to predicate an accordance with the form

$$D_1(Y_1, Y_2, Y_3) = \begin{cases} -1, & if \quad Y < Y_1 \\ 0, & if \quad Y = Y_2, \\ 1, & if \quad Y > Y_3 \end{cases} \qquad (10.12)$$

where the values Y_1, Y_2, Y_3 are accepted as earlier, that can be measured and used as the standard [32]. Consider zero as only one standard values for Y_1, Y_2, Y_3 it means that $Y_1 = Y_2 = Y_3 = 0$. Suppose that this rule holds true not only for the value of physical quantity, but also it includes its derivatives. We introduce the three types of variables: components of vector-indicator, which are formed from the processing of function and its derivatives, first and second order are formed by comparator (10.12), in which all standards are zero. Generalized our task and consider the function $L(\bar{x})$ was determined and n denotes the number of times of differentiable on interval of definition argument, $\bar{x}$. The expansion of $L(\bar{x} + \Delta\bar{x})$ is the scalar function of a vector argument in Taylor's series may be represented as a sum of terms in the form of product function $L(\bar{x})$ or $\Delta\bar{x}$– change of argument and components of vector is an indicator of variables includes partial derivatives

$$L(\bar{x}_p + \Delta\bar{x}_p) = L(\bar{x}_p)V1 + \Delta\bar{x}_p^T \begin{bmatrix} b_1 \\ \cdots \\ b_i \\ \cdots \\ b_n \end{bmatrix} + \Delta\bar{x}_p^T \|c_{ij}\| \frac{\Delta\bar{x}_p}{2}, \qquad (10.13)$$

where the elements of the corresponding constant and matrices are determined as follows

$$b_i = \left| \frac{\partial L\left(\bar{x}_n\right)}{\partial x_i} \right| V2_i; \quad c_{ji} = \left| \frac{\partial^2 L\left(\bar{x}_n\right)}{\partial x_j \partial x_i} \right| V3_{ji} \qquad (10.14)$$

$$V1 = D\left[L\left(\bar{x}_n\right)\right]; \quad V2_i = D\left[\frac{\partial L\left(\bar{x}_n\right)}{\partial x_i}\right]; \quad V3_{ji} = D\left[\frac{\partial^2 L\left(\bar{x}_n\right)}{\partial x_j \partial x_i}\right].$$

$$(10.15)$$

The developments of recurrent artificial neural network (RANN) for process-ing data on the basis of proposed instrument (10.12)–(10.15) are effective for scalar- and vector-function [32, 33]. Suppose that all components of vector-function $L(\bar{x} + \Delta\bar{x})$ of a vector argument $\bar{x}$ is three-time differentiable function we are setting the auxiliary variables

$$V1_i = D\left[L_i\left(\bar{x}_n\right)\right]; \qquad a_i = |L_i\left(\bar{x}_n\right)| V1_i$$

$$b_{ij} = \left| \frac{\partial L_i\left(\bar{x}_n\right)}{\partial x_j} \right| V2_{ij}; \quad c_{ij} = \left| \frac{\partial^2 L_i\left(\bar{x}_n\right)}{\partial x_j \partial x_i} \right| V3_{ij} \qquad (10.16)$$

where $L(\bar{x} + \Delta\bar{x})$ is the vector-function of a vector argument in a Taylor's series in the form of product quantities and corresponding components of vector-indicator including partial derivatives

$$L(\bar{x}_p + \Delta\bar{x}_p) = \begin{bmatrix} a_1 \\ \cdots \\ a_i \\ \cdots \\ a_n \end{bmatrix} + \|b_{ij}\| \Delta\bar{x}_p + \Delta\bar{x}_p^T \|c_{ij}\| \frac{\Delta\bar{x}_p}{2}. \qquad (10.17)$$

Thus, these new additional variables: components of vectors indicators are introduced by the Equations (10.15) and (10.16) are used in proposed new form of functions approximation. They are received after pre-processing by comparator (10.12) and can be determined to the form (10.13) or (10.17) through the elements of recurrent sequences. The expediency of introduction of vector-indicator was demonstrated for the search of root problem [32]. As it is shown in Reference [32] for scalar-function of a scalar's argument that these new variables are effective tool for creation of productive rules for control of System Support Decision Making (SSDM). In the case, when local maximum and minimum is presence into the range of consideration function, these tools are demonstrated of its efficiency and convergence. Particularly, taking into account the new expressions for the components of the vector-function of vector argument (10.17), it is important for developments system of control with prediction of vector-function was realized on the basis of RANN [41, 42].

10.4.2 Application of RANN and Problem of Analytic Learning for Neural Network

Let us consider ANN, which contains n inputs and p outputs. Assume that all inputs are given to neuron of first layer from the n sensors and dials initial conditions input n component of the vector $\bar{X}$. The network contains Q layers with N^q neurons in each layer. Each neuron from q layer has inputs and outputs. Two neurons: first i in direction propagation of information over the course from $q-1$ layer was connected to the second neuron j from q layer and this connection was characterized by synaptic weight coefficients (SWC) $\omega_{ij}^{(q)}$ and value of shift $\omega_{0j}^{(q)}$. Suppose that for M experiments, information about the values of all components of the input and output vectors are collected. We set the problem to find the value of SWC $\omega_{ij}^{(q)}$ that ensure the full compliance with the standards or satisfy the requirement of a minimum sum of squared standard deviations. We introduce n dimensional space, that appears in the input vector $\bar{X}$ and output vector $\bar{Y}$. Output from ith neuron is the input to jth neuron, as a result in the mth experiment, S_{mj}^q becomes jth neuron as input value and its activation function $h_{ij}^{(q)}(S_{mj}^q)$ is equal to the output $Y_{mj}^{(q)}$:

$$S_{mj}^q = \sum_{i=1}^{N^{(q)}} x_{mij}^{(q)}\omega_{ij}^{(q)} + \omega_{0j}^{(q)} = \sum_{i=1}^{N^{(q)}} y_{mji}^{(q-1)}\omega_{ij}^{(q)} + \omega_{0j}^{(q)}; \quad Y_{mj}^{(q)} = h_{ij}^{(q)}(S_{mj}^q);$$
$$D_{mjp} = Y_{mjp}^{(Q)},$$

where p is the pth components of $P^{(q)}$ dimensional output vector. For given designations, the analytical task of learning neural network was reduced to two tasks to ensure zero error or the problem for minimizing the sum of squares of errors

$$\min_{\omega} \sum_{m=1}^{M} \left[D_{mjp} - h_{ij}^{(q)}\left(S_{mj}^q\right) \right]^2; \quad j = \overline{1, N^q}; \quad p = \overline{1, P},$$

or combined task – minimizing the squared error and zero error for selected standard, whose solution was presented for the first time in Reference [17].

10.4.3 Modeling and Convergence of a Sequence of Synaptic Weight Coefficients (SWC)

In general, analytical solutions of this tasks was done by Trunov [23, 33], but from practical point of view its more interesting to receive the solution for calculations. For this purpose we shall be considering the task of minimizing the sum of squares of error for neuron with K inputs and one output, then the equations can be written as

$$
\begin{cases}
\sum\limits_{m=1}^{M} \left(Y_m - \dfrac{1}{1 + e^{-S_{mj}}} \right) \dfrac{e^{-S_{mj}}}{\left(1 + e^{-S_{mj}}\right)^2} = 0; & i = \overline{1, N^{q-1}}; \\[3mm]
\sum\limits_{m=1}^{M} \left(Y_m - \dfrac{1}{1 + e^{-S_{mj}}} \right) \dfrac{x_{mij} e^{-S_{mj}}}{\left(1 + e^{-S_{mj}}\right)^2} = 0; & j = \overline{1, N^q}; \\[3mm]
S_{mj} = S_{mj}^q = \sum\limits_{i=1}^{K} x_{mij}^{(q)} \omega_{ij}^{(q)} + \omega_{0ij}^{(q)} = \sum\limits_{i=1}^{K} y_{mij}^{(q-1)} \omega_{ij}^{(q)} + \omega_{0ij}^{(q)};
\end{cases}
\tag{10.18}
$$

For example, demonstration of convergence has been taken for neuron with one input and one output. For this case system (10.18) will transform to

$$
\begin{cases}
B_1 - \Delta\omega_{0n} A_{11} - \Delta\omega_{1n} A_{12} = 0 \\
B_2 - \Delta\omega_{0n} A_{21} - \Delta\omega_{1n} A_{22} = 0,
\end{cases}
$$

with solution which obtained as recurrence sequence

$$
\begin{cases}
\omega_{0n+1} = \omega_{0n} + \dfrac{A_{22} B_1 - A_{12} B_2}{A_{11} A_{22} - A_{21} A_{12}}; \\[3mm]
\omega_{1n+1} = \omega_{1n} + \dfrac{A_{21} B_1 - A_{11} B_2}{A_{12} A_{21} - A_{22} A_{11}}
\end{cases}
\tag{10.19}
$$

where $S_i = S_{mj}$ and constants are represented by using these notations as follows

$$
A_{11} = \sum_{i=1}^{M} \left\{ \frac{e^{-S_i} Y_i \left[1 - e^{-S_i}\right]}{\left(1 + e^{-S_i}\right)^3} \right\} - \sum_{i=1}^{M} \left\{ \frac{e^{-S_i} \left[1 - 2e^{-S_i}\right]}{\left(1 + e^{-S_i}\right)^4} \right\};
$$

$$
A_{12} = \sum_{i=1}^{M} \left\{ \frac{x_{1i} Y_i e^{-S_i} \left[1 - e^{-S_i}\right]}{\left(1 + e^{-S_i}\right)^3} \right\} - \sum_{i=1}^{M} \left\{ \frac{x_{1i} e^{-S_i} \left[1 - 2e^{-S_i}\right]}{\left(1 + e^{-S_i}\right)^4} \right\};
$$

$$A_{21} = \sum_{i=1}^{M} \left\{ \frac{x_{1i}e^{-S_i}Y_i\left[1 - e^{-S_i}\right]}{(1 + e^{-S_i})^3} \right\} - \sum_{i=1}^{M} \left\{ \frac{x_{1i} \ e^{-S_i}\left[1 - 2e^{-S_i}\right]}{(1 + e^{-S_i})^4} \right\};$$

$$(10.20)$$

$$A_{22} = \sum_{i=1}^{M} \left(\frac{x_{1i}^2 Y_i e^{-S_i}\left[1 - e^{-S_i}\right]}{(1 + e^{-S_i})^3} \right) - \sum_{i=1}^{M} \left\{ \frac{x_{1i}^2 e^{-S_i}\left[1 - 2e^{-S_i}\right]}{(1 + e^{-S_i})^4} \right\};$$

$$B_1 = \sum_{i=1}^{M} \left(\frac{Y_i e^{-S_i}}{(1 + e^{-S_i})^2} \right) - \sum_{i=1}^{M} \left(\frac{e^{-S_i}}{(1 + e^{-S_i})^3} \right);$$

$$B_2 = \sum_{i=1}^{M} \left(x_{1i}\frac{Y_i e^{-S_i}}{(1 + e^{-S_i})^2} \right) - \sum_{i=1}^{M} \left(x_{1i}\frac{e^{-S_i}}{(1 + e^{-S_i})^3} \right).$$

In general, an application of neural networks for control of physiotherapy process and particularly in phototherapy requires to make a selection from a number of examples, which has been successfully realized. Specificity of features for each practical example of prevention and treatment makes it impracticable for generalization and application for training, especially in the adaptive management in real time. Examples of analytical studies or processing of present data [23, 33], could be obtained due to comparative analysis of the present system behavior and bio-tissue patient [20–22] or controllable disturbance [42] or sequential controllable disturbances [41] with subsequent analysis of the reaction of system and changes in deviation of error. The implementation of the type being offered for the development of equipment to prevent failures, or using the tool for eliminating the failure and prevention, or algorithm localization of failures, are successfully realized and described [30]. However, the simultaneous use to analyze the behavior was not only for physical quantities but also for their derivatives and indicators [30, 32, 33], which was set by the comparator of type (10.12) to create a new opportunities for the application of ANN. Training and search of SWC in its essence was reduced to the problem of search the roots of nonlinear algebraic equations system [23, 33]. Under these conditions, the use of recurrent networks with memory and structural elements that determine the components of vector-indicator also requires effective tool for analytical

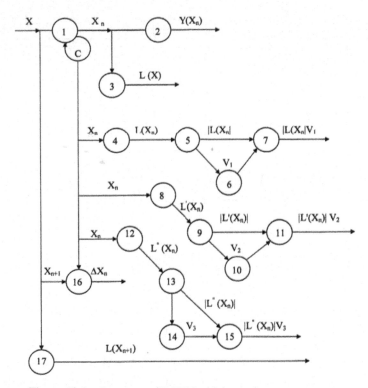

Figure 10.1   Fragment of RANN with long–short term memory.

learning or determination of roots. Figure 10.1 was demonstrated the fragment of such RANN, generalizing the idea of a new modification of Taylor's expression and its practical application and the implementation of recurrent parallel signal processing.

Thus, if the output of the neuron 3 defines the standard behavior of any system with any vector strategies $\bar{X}$, and from neurons 4 to 17 for $\bar{X}_n$ and $\bar{X}_{n+1}$, then decline

$$\Delta L(\bar{X}_{n+1}) = L(\bar{X}_{n+1}) - L(\bar{X}_n). \tag{10.21}$$

The strategy of change of control actions determines in accordance with the values of the components of the vector-indicator, which was obtained from vector-indicator component (10.6) or (10.11) or (10.15) are processed by comparator was described by Equation (10.12). Moreover, after adding the output of the neuron 7 with the output of the neuron 11 and add the product of the transposed vector with the output of the neuron 16 and the output of neuron 15 and multiplied by half of the vector output of neuron 16 can get approximated value as it is shown by Equation (10.17) at the point $\bar{X}_{n+1}$

$$L(\bar{X}_{n+1}) = \|a_i\| + \|b_{ij}\|\,\Delta\bar{X}_n + \Delta\bar{X}_n^T\,\|c_{ij}\|\,\frac{\Delta\bar{X}_n}{2}. \qquad (10.22)$$

In turn, this standard element (Figure 10.1) also makes it possible to estimate the approximation error by comparing the values $L(\bar{X}_{n+1})$ was obtained by Equation (10.22) and output (10.17). The latter, in turn, opens the possibility to carry out improvement model and formation of control rules. This is especially true for real systems in which $L(\bar{X}_{n+1})$ is oscillating and non-smooth functions, but for such cases it requires a lot of point approximations and expanded the volume of random memory.

As it was shown in Reference [32] for non-smooth-oscillating scalar's functions in which the value x_1, for which vector-indicator is equal to $(1, -1, \ 1)$ that for any schemes of approximations was received for the next value x_2, for which vector-indicator is equal to $(1, 0, 1)$ or $(1, 1, 0)$ then the process of recurrent approximation need to be interrupted. Consider the order of action in accordance with the rule:

1. Accept $L(x_3) = L(x_2)$;
2. Determine the decomposition in series

$$L(x_3) = L(x_2) + L'(x_2)\Delta + L''(x_2)\frac{\Delta^2}{2},$$

where we notice "−" as an increment:

$$\Delta = -\frac{2L'(x_2)}{L''(x_2)};$$

3. Check the conditions, if $\forall x \in [x_2, x_3] L(x) \geq L(x_2) = L(x_3)$ and $L''(x_2) \neq 0$, than x_3 the new approach can be defined independently of the features of the function behavior on this interval as

$$x_3 = x_2 - \frac{2L'(x_2)}{L''(x_2)}.$$

As it follows from this consideration, if the result of the functions properties and due to the choice of approximation of the initial point, one of the approximation falls within the range in which there is a loop, the loop process is unable to anticipate and eliminate. The latter was carried out by applying the above rules. To summarize, we introduce the control function, W. It determines the value of the increment of the first and second components and the value of the third components of the vector-indicator, then the present expansion can be written as

$$L(x_2) + L'(x_2)\Delta + L''(x_2)\frac{\Delta^2}{2} = L(x_3)W, \qquad (10.23)$$

where

$$L(x_3)W = \begin{cases} L(x_2), & \text{if } \Delta V_1 = 0, \text{ and } |\Delta V_2| \geq 1 \text{ and } V_3 \neq 0 \\ 0 \end{cases}.$$

The correctness of conditions (10.23) observes only under condition $\Delta V_1 = 0$ and under additional conditions: increment of second components, processed by comparator (12) equals unity, $\Delta V_2 = 1$; third components, it is obligatory does not equal to zero, i.e., $V_3 \neq 0$. This approach can be generalized for three serial point's scheme x_{n-1}, x_n, x_{n+1} by the method of recurrent approximation [30, 32]:

$$x_{n+1} = x_{n-1} - \frac{|L(x_{n-1})| V_{1n-1} V_{2n-1}}{|L'(x_{n-1})|} - \frac{|L(x_n)|}{2|L'(x_n)|} \left| \left(V_{1n-1} + \frac{\Delta V_{1n}}{2} \right) \right.$$

$$\left. \left(V_{2n-1} + \frac{\Delta V_{2n}}{2} \right) \right| - \frac{|L'(x_n)| V_{1n+1}}{|L''(x_n)|} \Delta V_{2n+1}. \qquad (10.24)$$

Thus, the management of strategies for the search process of irradiation region, intensity and dose with defined level of parameters can be simplified and reduced to the search for the roots of its differences. Proposed methodology was based on recurrent approximation, vector-indicator, RANN with analytic learning after the formation of appropriate set of rules can be used for design of SSDM in physiotherapy. Construction of additional logical rules of data processing eliminates the loop and improves the search algorithm of roots, even when the function is nonlinear and non-smooth.

10.5 Statement and Solution of the Control Efficiency Problem During Physiotherapy Process

Let us assume that the description of process of functioning of equipment and its components for physiotherapy during early diagnostics, prevention and treatment and ancillary transactions was generalized by means of nonlinear mathematical models:

$$\frac{d}{dt}\bar{\Psi} = f(\bar{X}(t), t, [A], [C], [I], \bar{U}), \qquad (10.25)$$

where $\bar{X}(t)$ is the n-component of vector strategy, which is represent as independent function of time, t, $[A]$ is the kinematics matrix of the kinematics parameters of the equipment are used during the independent motion of three axis or the manipulator for motion of module for controlling and shifting to spatial position and is introduced by Trunov [30], $[C]$ and $[I]$ are the matrix of resistance coefficients and added mass and moments of inertia, $\bar{U}$ is the vector control action. Suppose, the description of the standard behavior for manipulator and its systems are done by the vector-function of vector strategies $\bar{X}_e(t)$, simultaneously with the function of purpose

$$\bar{\Psi}(t) = \bar{X}(t) - \bar{X}_e(t). \tag{10.26}$$

Defines the value of efficiency as well k-dimensional vector whose components are calculated according to the methodology [33] and vector-functions and management strategies, as well as the properties of the object and external influences

$$Q = f_1(\bar{Y}, \bar{X}, [A], [C], [I], \bar{U}).$$

Let us denoted

- vector-function of efficiency deviation

$$\delta = (Q - Q^*)/\|Q^*\|; \tag{10.27}$$

- objective function

$$F(\bar{Y}, \bar{X}, [A], [C], [I], \bar{U}) = \frac{1}{2}[\delta]^T [P][\delta], \tag{10.28}$$

where $[P]$ is a positively defined matrix, other properties of which does not determined with norm of standard behavior description of system on interval of time:

$$\|Q^*\| = \sqrt{\int_0^t f_1(\bar{Y}, \bar{X}_e, [A], [C], [I], \bar{U})^2 dt}.$$

10.5.1 Pointing the Problem of Minimizing the Objective Function

$$\min_{\bar{U}} \left\{ \frac{1}{2}[\delta]^T [P][\delta] \right\},$$

with constraint inequality

$$g_j = f_{3j}(\bar{Y}, \bar{X}, [A], [C], [I], \bar{U}) - b_j \leq 0; \quad j = \overline{1, m}, \tag{10.29}$$

for this case, the Lagrange's function can be written as follows

$$L(\bar{X}, \bar{\Lambda}) = \frac{1}{2}[\delta]^T[P][\delta] + \sum_{j=1}^{m} \lambda_j (f_{3j}(\bar{Y}, \bar{X}, [A], [C], [I], \bar{U}) - b_j). \quad (10.30)$$

Introducing a vector-function, $\bar{F}_3(\bar{Y}, \bar{X}, [A], [C], [I], \bar{U})$, components of which are being equal the function $f_{3j}(\bar{Y}, \bar{X}, [A], [C], [I], \bar{U})$ and column matrix $[b]$ with components b_j, then Lagrange function can be rewritten as

$$L(\bar{X}, \bar{\Lambda}) = \frac{1}{2}[\delta]^T[P][\delta] + \bar{\Lambda}^T[\bar{F}_3(\bar{Y}, \bar{X}, [A], [C], [I], \bar{U}) - [b]]. \quad (10.31)$$

Then in accordance with the method of Lagrange multipliers write the system for solution any $\bar{X}$ is a vector strategies and $\bar{\Lambda}$ is a vector of Lagrange's multipliers, $\bar{V}$ and $\bar{W}$ are additional vectors are satisfied the conditions

$$\begin{cases} \frac{1}{2}\nabla_x([\delta]^T[P][\delta]) + \\ \quad + \bar{\Lambda}^T \nabla_x \left\{ [\bar{F}_3(\bar{Y}, \bar{X}, [A], [C], [I], \bar{U}) - [b]]^T \right\} - \bar{V} = 0; \\ [\bar{F}_3(\bar{Y}, \bar{X}, [A], [C], [I], \bar{U}) - [b]] + \bar{W} = 0; \\ \bar{X}^T \bar{V} = 0; \\ \bar{\Lambda}^T \bar{W} = 0. \end{cases} \quad (10.32)$$

The first and second equation of system (32) are multiplied accordingly to the vectors strategies and Lagrange multipliers, which have been transposed after subtraction, then the product of second equation and the product of first equation taking into account for the third and fourth equations and linearity of constraints with saddle point conditions $\Delta \bar{X}^T \nabla_x([\delta]^T[P][\delta]) = 0$, therefore we conclude

$$\sum_{j=1}^{m} \lambda_j b_j = 0. \quad (10.33)$$

Taking into account the complementary slackness conditions and by comparing it with last results have been find the value of utility vector $\bar{W}$

$$\bar{W} = [b_1, \ldots, b_i, \ldots, b_n]^T.$$

Moreover, in addition under existence the physical meaning of Lagrange multipliers could be written as

$$\lambda_j = \operatorname*{root}_{\lambda_j} \left\{ \sum_{j=1}^{m} \lambda_j \left[\nabla_x f_{3j}(\bar{Y}, \bar{X}, [A], [C], [I], \bar{U}) \right] + \right.$$
$$\left. + \frac{1}{2} \left[\nabla_x \left([\delta]^T [P][\delta] \right) \right] = 0 \right\},$$

that includes additional equation

$$\sum_{j=1}^{m} \left\{ \operatorname*{root}_{\lambda_j} \left\{ \sum_{j=1}^{m} \lambda_j \left[\nabla_x f_{3j} (\bar{Y}, \bar{X}, [A], [C], [I], \bar{U}) \right] + \right. \right.$$
$$\left. \left. + \frac{1}{2} \left[\nabla_x \left([\delta]^T [P][\delta] \right) \right] = 0 \right\} \right\} b_j = 0, \qquad (10.34)$$

is the complement system (10.32). Homogeneity of Equation (10.34) explains that the solution of system does not provide a unified solution. Taking into account newly additional Equation (10.34) and the system (10.32) is rewritten in the form

$$\begin{cases} \frac{1}{2} \nabla_x \left([\delta]^T [P][\delta] \right) + \nabla_x \sum_{j=1}^{m} \lambda_j f_{3j} (\bar{Y}, \bar{X}, [A], [C], [I], \bar{U}) - \bar{V} = 0; \\ \nabla_\lambda \sum_{j=1}^{m} \lambda_j \left(f_{3j} (\bar{Y}, \bar{X}, [A], [C], [I], \bar{U}) - b_j \right) + [b] = 0; \\ \sum_{j=1}^{m} \operatorname*{root}_{\lambda_j} \left\{ \nabla_x \left([\delta]^T [P][\delta] \right) \left[\nabla_x f_{3j} (\bar{Y}, \bar{X}, [A], [C], [I], \bar{U}) \right]^{-1} \right\} b_j = 0; \\ \bar{X}^T \bar{V} = 0; \\ \bar{\Lambda}^T [b] = 0. \end{cases}$$

The latter system is not complete, so as a result, the solution for this system is not single and enhanced the degree of completeness has been used as the result of work, which are partly highlighted in Reference [30]. If we apply the objective function in a form of recurrent approach, then for $n+1$th approach we get an approximation

$$F\left(\bar{X}_{n+1}\right) = \frac{1}{2}\left\{[\delta]^T[P][\delta]\right\}\Big|_{\bar{X}=\bar{X}_n} + \frac{1}{2}\Delta\bar{X}_n^T\left\{\nabla_x\left([\delta]^T[P][\delta]\right)\right\}\Big|_{\bar{X}=\bar{X}_n} +$$

$$+ \frac{1}{4}\Delta\bar{X}_n^T\left\{\nabla_x^2\left([\delta]^T[P][\delta]\right)\right\}\Big|_{\bar{X}=\bar{X}_n}\Delta\bar{X}_n$$

Substituting a new expression of objective function in Equation (10.31) gives Lagrange function

$$L\left(\bar{X},\bar{\Lambda}\right) = \frac{1}{2}\left\{[\delta]^T[P][\delta]\right\}\Big|_{\bar{X}=\bar{X}_n} + \frac{1}{2}\Delta\bar{X}_n^T\left\{\nabla_x\left([\delta]^T[P][\delta]\right)\right\}\Big|_{\bar{X}=\bar{X}_n} +$$

$$+ \frac{1}{4}\Delta\bar{X}_n^T\left\{\nabla_x^2\left([\delta]^T[P][\delta]\right)\right\}\Big|_{\bar{X}=\bar{X}_n}\Delta\bar{X}_n +$$

$$+ \bar{\Lambda}^T\left[\bar{F}_3\left(\bar{Y},\bar{X},[A],[C],[I],\bar{U}\right) - [b]\right],$$

then disclosing the selected objective function and express the gradient in first and second order, after simple transformations, so, we write

$$\nabla_x\left([\delta]^T[P][\delta]\right) = \nabla_x\left([\delta]^T[P]\right)[\delta] + [\delta]^T[P]\nabla_x([\delta]);$$

$$\nabla_x^2\left([\delta]^T[P][\delta]\right) = \nabla_x^2\left([\delta]^T[P]\right)[\delta] + 2\nabla_x\left([\delta]^T[P]\right)\nabla_x([\delta]) +$$

$$+ [\delta]^T[P]\nabla_x^2([\delta]).$$

For this assumption we find that the auxiliary vector $\bar{V}$ depends on the properties of objective function, namely

$$\bar{V} = -\frac{1}{2}\left\{\nabla_x^2\left([\delta]^T[P][\delta]\right)\right\}\Big|_{\bar{X}=\bar{X}_n}\Delta\bar{X}_n =$$

$$= -\frac{1}{2}\left\{\begin{array}{l}\nabla_x^2\left([\delta]^T[P]\right)[\delta] + \\ + 2\nabla_x\left([\delta]^T[P]\right)\nabla_x([\delta]) + [\delta]^T[P]\nabla_x^2([\delta])\end{array}\right\}\Bigg|_{\bar{X}=\bar{X}_n}\Delta\bar{X}_n.$$

The second possible solution is, when the point of solution is not a saddle point, then the condition

$$\frac{1}{2}\Delta\bar{X}^T\nabla_x\left([\delta]^T[P][\delta]\right) + \sum_{j=1}^m \lambda_j b_j = 0 \quad \text{or}$$

$$\frac{1}{2}\Delta\bar{X}^T\nabla_x\left([\delta]^T[P][\delta]\right) + \bar{\Lambda}^T[b] = 0,$$

then the system can be supplied with an increased number of equations, but with an unknown auxiliary vector $\bar{W}$

$$
\begin{cases}
\frac{1}{2}\nabla_x\left([\delta]^T[P][\delta]\right) + \nabla_x\left\{\sum_{j=1}^m \lambda_j f_{3j}\left(\bar{Y}, \bar{X}, [A], [C], [I], \bar{U}\right)\right\} - \bar{V} = 0; \\[2mm]
\bar{V} = -\frac{1}{2}\left\{\nabla_x^2\left([\delta]^T[P][\delta]\right)\right\}\Big|_{\bar{X}=\bar{X}_n}\Delta\bar{X}_n; \\[2mm]
\frac{1}{2}\Delta\bar{X}^T\nabla_x\left([\delta]^T[P][\delta]\right) + \bar{\Lambda}^T[b] = 0; \\[2mm]
\nabla_\lambda\left\{\sum_{j=1}^m \lambda_j\left(f_{3j}\left(\bar{Y}, \bar{X}, [A], [C], [I], \bar{U}\right) - b_j\right)\right\} + \bar{W} = 0; \\[2mm]
\sum_{j=1}^m\left\{\operatorname*{root}_{\lambda_j}\left\{\sum_{j=1}^m \lambda_j\left[\nabla_x f_{3j}\left(\bar{Y}, \bar{X}, [A], [C], [I], \bar{U}\right)\right] + \right. \right. \\[2mm]
\qquad\qquad\left.\left. +\frac{1}{2}\left[\nabla_x\left([\delta]^T[P][\delta]\right)\right] = 0\right\}\right\}b_j = 0; \\[2mm]
\bar{X}^T\bar{V} = 0; \quad \bar{\Lambda}^T\bar{W} = 0; \quad \bar{X}_n = \bar{X}_{n-1} + \Delta\bar{X}_n.
\end{cases}
$$

(10.35)

However, to make an assumption, that the result of the deviation of state, a new value of objective function reaches to zero, in other words – deviation of efficiency from standard value reaches to zero. From this assumption, we get the expression from the first equation of last system on the basis of the objective function (10.28)

$$
\frac{1}{2}\Delta\bar{X}^T\nabla_x\left\{[\delta]^T[P][\delta]\right\} = -\Delta\bar{X}^T\sum_{j=1}^m\lambda_j\nabla_x f_{3j}\left(\bar{Y}, \bar{X}, [A], [C], [I], \bar{U}\right).
$$

(10.36)

Assume that the search strategy of incorporated management ensuring that the specified value vector of efficiency or maximum rate of change of the vector of efficiency, then the feeding rate of change of deviation efficiency as derivative complex function, we write

$$
\begin{aligned}
\frac{d[\delta]}{dt} &= \frac{\partial[\delta]}{\partial t} + \left[\nabla_x^T(\delta_i)\right]\frac{d}{dt}\left(\bar{\Psi} + \bar{X}_e\right) = \\
&= \frac{\partial[\delta]}{\partial t} + \left[\nabla_x^T(\delta_i)\right]\left[f\left(\bar{X}(t), t, [A], [C], [I], \bar{U}\right) + \frac{d}{dt}\bar{X}_e\right].
\end{aligned}
$$

Under the conditions of given value of speed deviation increase the efficiency, taking into account the properties of the object management model is

$$[\nabla_x^T (\delta_i)] = -\frac{\partial [\delta]}{\partial t} \left[f\left(\bar{X}(t), t, [A], [C], [I], \bar{U}\right) + \frac{d}{dt}\bar{X}_e \right]^{-1},$$

was indicated by upper index minus unity – inverse matrix and

$$[\nabla_x^T (\delta_i)] = \left[\nabla_x \left([\delta]^T\right)\right]^T = [a_{ij}] = \left[\frac{\partial \delta_i}{\partial x_j}\right]$$

is a matrix dimension $k \times n$. Under these conditions, the system is complemented by the equation and then enter the vector of efficiency with component in according to the methodology [31, 32], this system can be obtained

$$
\begin{cases}
\dfrac{1}{2}\nabla_x \left([\delta]^T [P][\delta]\right) + \nabla_x \left\{ \sum_{j=1}^{m} \lambda_j f_{3j}\left(\bar{Y}, \bar{X}, [A], [C], [I], \bar{U}\right) \right\} - \bar{V} = 0; \\[2mm]
\bar{V} = -\dfrac{1}{2}\left\{ \nabla_x^2 \left([\delta]^T [P][\delta]\right) \right\}\Big|_{\bar{X}=\bar{X}_n} \Delta \bar{X}_n; \quad \bar{X}_n = \bar{X}_{n-1} + \Delta \bar{X}_n; \\[2mm]
\dfrac{d[\delta]}{dt} = \dfrac{\partial[\delta]}{\partial t} + [\nabla_x^T (\delta_i)] \left(f\left(\bar{X}(t), t, [A], [C], [I], \bar{U}\right) + \dfrac{d}{dt}\bar{X}_e \right); \\[2mm]
\nabla_\lambda \left\{ \sum_{j=1}^{m} \lambda_j \left(f_{3j}\left(\bar{Y}, \bar{X}, [A], [C], [I], \bar{U}\right) - b_j \right) \right\} + [b] = 0; \\[2mm]
\sum_{j=1}^{m} \operatorname*{root}_{\lambda_j} \left\{ \sum_{j=1}^{m} \lambda_j \left[\nabla_x f_{3j}\left(\bar{Y}, \bar{X}, [A], [C], [I], \bar{U}\right)\right] + \right. \\[2mm]
\qquad \left. + \dfrac{1}{2}\left[\nabla_x \left([\delta]^T[P][\delta]\right)\right] = 0 \right\} b_j = 0; \\[2mm]
\Delta\bar{X}^T \nabla_x \left([\delta]^T[P][\delta]\right) + \sum_{j=1}^{m} \lambda_j b_j = 0; \quad \bar{X}^T \bar{V} = 0; \quad \bar{\Lambda}^T [b] = 0; \\[2mm]
Q_i = f_{1i}\left(\bar{Y}, \bar{X}, [A], [C], [I], \bar{U}\right); \\[2mm]
\delta_i \left(\bar{Y}, \bar{X}, [A], [C], [I], \bar{U}\right) = \dfrac{Q_i - Q_i^*}{\|Q^*\|}.
\end{cases}
$$

$$(10.37)$$

If, in addition, we set a goal: to provide the highest from range of possible velocity of change of deviation vector-function of efficiency, and in moment, when it reaches the design value and becomes a stable $\frac{d[\delta]}{dt} = 0$, at this moment it reaches to design values Q_i^* of the vector of relative effectiveness is $[\delta] = 0$, given that the expression of gradient of second order objective function (10.27), the first three equations of the system (10.37) will get the form regardless of the control algorithm:

$$\nabla_x \left\{ \sum_{j=1}^{m} \lambda_j f_{3j} \left(\bar{Y}, \bar{X}, [A], [C], [I], \bar{U} \right) \right\} + \frac{1}{2} \nabla_x \left([\delta]^T [P] \right) \nabla_x \left([\delta] \right) = 0.$$

The latter is true regardless of the reference system behavior conditions except when

$$\left(f \left(\bar{X}(t), t, [A], [C], [I], \bar{U} \right) + \frac{d}{dt} \bar{X}_e \right) = 0. \tag{10.38}$$

Suppose, that the number of components of the vector strategies n is more than or equal to the number of constraints m, then the solution of (10.35) determine the value of the Lagrange multipliers or in other words complement system becomes independent Equation (10.38) to search for optimal strategies with the prior synthesis of matrix $[P]$.

As an example of the formation of algorithm for control with optimal selection of vector strategies according to the constraints inequality and conditions of the maximum speed for decline deviation of vector efficiency with separate linear inputs take the system

$$\frac{d}{dt} \bar{\Psi} = f \left(\bar{X}(t), t, [A], [C], [I], \bar{U} \right) = G \left(\bar{X}(t), t, [A], [C], [I] \right) +$$

$$+ B \left(\bar{X}(t), t, [A], [C], [I] \right) \bar{U} \left(\bar{X}(t), t, [A], [C], [I] \right),$$

where $G \left(\bar{X}(t), t, [A], [C], [I] \right)$, $B \left(\bar{X}(t), t, [A], [C], [I] \right)$ – marked – n-measurable vector-functions and vector control $\bar{U} \left(\bar{X}(t), t, [A], [C], [I] \right)$. The last equation can be rewritten in a simplified form

$$\frac{d}{dt} \bar{\Psi} = G \left(\bar{X}(t), t \right) + B \left(\bar{X}(t), t \right) \bar{U} \left(\bar{X}(t), t \right),$$

and the substantial derivative under these conditions will become

$$\frac{d[\delta]}{dt} = \frac{\partial[\delta]}{\partial t} + [\nabla_x^T(\delta_i)]\left(G\left(\bar{X}(t),t\right) + B\left(\bar{X}(t),t\right)\bar{U}\left(\bar{X}(t),t\right) + \frac{d}{dt}\bar{X}_e\right),$$

where used allegation was proved by direct differentiation vector $\nabla_x([\delta]) = [\nabla_x^T(\delta_i)]$.

$$\begin{cases} \dfrac{1}{2}\nabla_x\left([\delta]^T[P][\delta]\right) + \nabla_x\left\{\sum_{j=1}^{m}\lambda_j f_{3j}\left(\bar{Y},\bar{X},[A],[C],[I],\bar{U}\right)\right\} - \bar{V} = 0; \\[4mm] \bar{V} = -\dfrac{1}{2}\left\{\nabla_x^2\left([\delta]^T[P][\delta]\right)\right\}\Big|_{\bar{X}=\bar{X}_n}\Delta\bar{X}_n; \quad \bar{X}_n = \Delta\bar{X}_n; \\[4mm] \nabla_\lambda\left\{\sum_{j=1}^{m}\lambda_j\left(f_{3j}\left(\bar{Y},\bar{X},[A],[C],[I],\bar{U}\right) - b_j\right)\right\} + \bar{W} = 0; \\[4mm] \Delta\bar{X}^T\nabla_x\left([\delta]^T[P][\delta]\right) + \sum_{j=1}^{m}\lambda_j b_j = 0; \quad \bar{X}^T\bar{V} = 0; \quad \bar{\Lambda}^T[b] = 0; \\[4mm] Q_i = f_{1i}\left(\bar{Y},\bar{X},[A],[C],[I],\bar{U}\right); \delta_i\left(\bar{Y},\bar{X},[A],[C],[I],\bar{U}\right) = \dfrac{Q_i - Q_i^*}{\|Q\|}; \end{cases}$$

Now let us consider another approach for solving the problems and obtain substantial derivative for the objective function

$$\begin{aligned} \frac{d[F]}{dt} &= \frac{\partial[F]}{\partial t} + \nabla_x([F])\frac{d}{dt}(\bar{\Psi} + \bar{X}_e) = \\[2mm] &= \frac{\partial[F]}{\partial t} + \nabla_x([F])\left(f(\bar{X}(t),t,[A],[C],[I],\bar{U}) + \frac{d}{dt}\bar{X}_e\right) = \\[2mm] &= \frac{\partial[F]}{\partial t} + \nabla_x([F])\left(G(\bar{X}(t),t) + B(\bar{X}(t),t)\bar{U}(\bar{X}(t),t) + \frac{d}{dt}\bar{X}_e\right) \end{aligned}$$

here the goal is to reduce the size of recording symbols which is conditionally marked

$$F = F(\bar{Y},\bar{X},[A],[C],[I],\bar{U})$$

and reformulate the system in next form

$$
\begin{cases}
\frac{1}{2}\nabla_x([\delta]^T[P][\delta]) + \sum_{j=1}^{m} \lambda_j \nabla_x[f_{3j}(\bar{Y}, \bar{X}, [A], [C], [I], \bar{U})] - \bar{V} = 0; \\[2mm]
\bar{V} = -\frac{1}{2}\left\{\nabla_x^2([\delta]^T[P][\delta])\right\}\Big|_{\bar{X}=\bar{X}_n} \Delta\bar{X}_n; \quad \bar{X}_n = \bar{X}_{n-1} + \Delta\bar{X}_n; \\[2mm]
\frac{d[F]}{dt} = \frac{\partial[F]}{\partial t} + \nabla_x([F])\left(G(\bar{X}(t), t) + B(\bar{X}(t), t)\bar{U}(\bar{X}(t), t) + \frac{d}{dt}\bar{X}_e\right); \\[2mm]
\nabla_\lambda\left\{\sum_{j=1}^{m} \lambda_j(f_{3j}(\bar{Y}, \bar{X}, [A], [C], [I], \bar{U}) - b_j)\right\} + [b] = 0; \\[2mm]
\frac{1}{2}\Delta\bar{X}^T \nabla_x([\delta]^T[P][\delta]) + \sum_{j=1}^{m} \lambda_j b_j = 0; \quad \bar{X}^T\bar{V} = 0; \quad \bar{\Lambda}^T[b] = 0; \\[2mm]
Q_i = f_{1i}(\bar{Y}, \bar{X}, [A], [C], [I], \bar{U}); \\[2mm]
\delta_i(\bar{Y}, \bar{X}(t), t, [A], [C], [I], \bar{U}) = \dfrac{Q_i(\bar{Y}, \bar{X}(t), t, [A], [C], [I], \bar{U}) - Q_i^*}{\|Q^*\|}.
\end{cases}
$$

For the synthesis parameters of control with assembled components of recurrent networks with memory of states will submit the growth of state vector:

$$
\Delta\bar{X} = \frac{d}{dt}(\bar{X})\Delta t = \frac{d}{dt}(\bar{\Psi} + \bar{X}_e)\Delta t =
$$

$$
= [f(\bar{X}(t), t, [A], [C], [I], \bar{U}) + \frac{d}{dt}\bar{X}_e]\Delta t =
$$

$$
= [G(\bar{X}(t), t) + B(\bar{X}(t), t)\bar{U}(\bar{X}(t), t) + \frac{d}{dt}\bar{X}_e]\Delta t;
$$

$$
\bar{X}_n = \bar{X}_{n-1} + \Delta\bar{X}_n = [G(\bar{X}(t), t) + B(\bar{X}(t), t)\bar{U}(\bar{X}(t), t) +
$$

$$
+ \frac{d}{dt}\bar{X}_e](T_n - T_{n-1}).
$$

Therefore, subject to the availability of the results of early diagnosis, and medical prescriptions are formed as the functions of space coordinates for such variables as: areas of exposure; intensity and exposure dose and time. For this data are formed by the components of deviation of efficiency vector and found the solution of system (10.37) such quantities as: vector of states; vector of Lagrange multipliers.

10.6 Modeling and Convergence of a Sequence of SWC

Generic neurons, analytically trained, allows to create a system of physical rehabilitation with elements of diagnostics. It's structure includes camera and components for early diagnosis sensor with cameras, electrocardiograph, pulse oxygen meter, actuators RANN, system of physiotherapy impact and control. The fragment of scheme as one from the units that are investigated was shown in Figure 10.2. Implementation of the principles of resonance conformational therapy are proposed by forming a combined ANN and RANN photon as a part of SSDM [11–13]. New diagnostic and control tools are

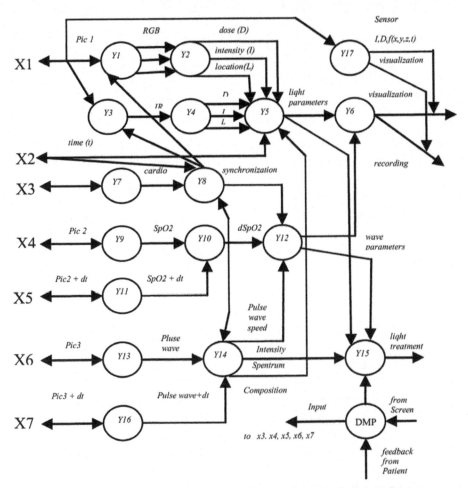

Figure 10.2 Schematic view of SSDM fragment for control of phototherapy.

applied for increasing the efficiency and for satisfaction of demand automation of the process of diagnosis, prevention and treatment [11, 12]. To determine the state of patient experimentally are measured by the spectral of composition of the radiation in the reflected and passing rays, simultaneously, analyzing the cardio signals are implemented from mobile ECG. The fixation value of the local pulse in ECG diagram at the moment of time t (pic. 1, pic. 2, pic. 3 of Figure 10.2) and make measurements at the moment $t + dt$ – synchronously in two points, at certain referenced distances along the arteries, determine the wave velocity and the saturation of blood.

The structure of such system includes the diagnostic structural elements, RANN and ANN, which is completed from a set of standard single-layer neurons [23]. The analysis was based on synchronization of received signals has been formed controlling influences, for which was governed and provided the magnitude of intensity, dose, irradiation area and the velocity of propagation light bands in biological tissues. Feedback informations are: the oxygen concentration and the gloale maximum of the pulse wave has allow to control the process of phototherapy in real time. Proposed standards of neurons and process of analytic learning networks are obtained by recurrent sequences is in general feasible. Examples of numerical studies, which are demonstrated by the nature of SWC convergence and was defined by sequence (10.19), are given in Table 10.1. As show the results of numerical simulation are given in Tables 10.1–10.4, which are presented for different initial approaches values of SWC are rapidly converging. The magnitude and sign of initial approximation

Table 10.1 The values of SWC RANN for nth approximation by linear scheme

n	ω_{0n}	ω_{1n}	ω_{0n}	ω_{1n}
0	−0.8	0.8	−0.3	0.3
1	0.232549	2.45108	0.408325	1.641963
2	−0.06061	3.418428	0.107067	2.571902
3	−0.27358	4.390314	−0.13307	3.555473
4	−0.39053	5.267905	−0.3119	4.518139
5	−0.42281	5.948943	−0.40293	5.366699
6	−0.4243	6.391915	−0.42419	6.015852
7	−0.4233	6.584284	−0.42416	6.427866
8	−0.42312	6.616587	−0.42325	6.593449
9	−0.42311	6.6174	−0.42311	6.616972
10	−0.42311	6.6174	−0.42311	6.6174
11	−0.42311	6.6174	−0.42311	6.6174

Table 10.2   The values of SWC RANN for nth approximation by linear scheme

n	ω_{0n}	ω_{1n}	ω_{0n}	ω_{1n}
0	−0.5	0.5	−1	1
1	0.348296	1.918929	0.226463	2.736287
2	0.048179	2.86999	−0.09139	3.693219
3	−0.18609	3.853674	−0.30202	4.651896
4	−0.34588	4.794877	−0.40256	5.488551
5	−0.41371	5.590638	−0.42399	6.105785
6	−0.42476	6.170496	−0.4239	6.473803
7	−0.42376	6.503514	−0.42319	6.603236
8	−0.42376	6.503514	−0.42311	6.617249
9	−0.42316	6.608308	−0.42311	6.6174
10	−0.42316	6.608308	−0.42311	6.6174
11	−0.42311	6.617338	−0.42311	6.6174

Table 10.3   The values of SWC RANN for nth approximation

n	ω_0	ω_1	ω_0	ω_1
0	0.5	0.5	1	1
1	0.243522	1.506316	0.226463	2.736287
2	0.051021	2.453176	−0.09139	3.693219
3	−0.15255	3.439809	−0.30202	4.651896
4	−0.31466	4.403316	−0.40256	5.488551
5	−0.40187	5.262655	−0.42399	6.105785
6	−0.42418	5.937241	−0.4239	6.473803
7	−0.4244	6.384425	−0.42319	6.603236
8	−0.42331	6.582194	−0.42311	6.617249
9	−0.42312	6.616483	−0.42311	6.6174
10	−0.42311	6.617399	−0.42311	6.6174
11	−0.42311	6.6174	−0.42311	6.6174
12	−0.42311	6.6174	−0.42311	6.6174
13	−0.42311	6.6174	−0.42311	6.6174

of solution are quantities which are weakly effecting on the speed of convergence. In addition, it should be noted that used quadratic approximation scheme practically [33] does not reduce the number of iterations of approximations till its full convergence. The latter allows to claim: the quadratic approximation scheme is a bi-side and quadratic convergence; its speed mainly depends from the properties of activation functions and the type of

Table 10.4 The values of SWC RANN for nth approximation by linear scheme

n	ω_0	ω_1	ω_0	ω_1
0	0.5	5	–0.5	5
1	–0.6098	4.273501	–0.43524	5.641739
2	–0.45191	4.896313	–0.4261	6.190138
3	–0.43134	5.601353	–0.42375	6.511489
4	–0.42602	6.165245	–0.42316	6.609489
5	–0.42381	6.500538	–0.42311	6.617353
6	–0.42317	6.60784	–0.42311	6.6174
7	–0.42311	6.617331	–0.42311	6.6174
8	–0.42311	6.6174	–0.42311	6.6174
9	–0.42311	6.6174	–0.42311	6.6174
10	–0.42311	6.6174	–0.42311	6.6174
11	–0.42311	6.6174	–0.42311	6.6174
12	–0.42311	6.6174	–0.42311	6.6174

task and to a lesser degree from approximation schemes [33]. So for the aims of learning of neurons in which the predetermine value of error was determined by maximum value of derivatives three- or fourth-order of the original activation function, and for the problem of minimizing the sum of squares of errors by maximize the value of derivative of the fourth and fifth order. However, oscillation of operator appreciably make worse of SWC convergence. As demonstrate results of numerical experiments independently from the differences of first approximation sequence convergent to equal value with accuracy until fifth sign.

10.7 Conclusions

1. The mathematical modeling of EMW interaction with the electron of radical from terminal enzyme of respiratory chain of cytochrome-c-oxidase and other photo acceptors allows to determine.

   - the conditions, that increase effectiveness of radiation due to reducing the frequency of photons for activation of photo acceptors;
   - properties of proportional symmetry of two splitting line in secondary radiation spectrum composition of radiation which open the way to new algorithm for early diagnostics.

2. The Problem of Synthesis of Neural Network for Control of Processes Phototherapy was reduced by the means of the recurrent approximation

to solution of system of linear algebraic equations for SWC as recurrent sequence, which are represented in analytic form.

3. The rate of convergence of solutions are generated as proposed sequence was determined by evaluation of maximum values of derivatives and activation functions. It does not depend from the value of selected initial approximations, which creates prerequisites almost instantaneous learning neurons in analytical form.

4. The structure of the system control of phototherapy process was based on application of standardized elements of ANN and RANN realizing analytical learning algorithms can be used for synthesis of intelligent automatized control systems with simultaneous survey of several magnitude of error of intensity, dose, velocity of propagation light bands, number of electrons detachments from of mitochondrion of cells synchronized with cardio signal of patient and light treatment.

5. Introduction of vector-indicator of physical quantity its speed and acceleration allows to expand continual vector-function of vector argument and formed RANN for data processing and create algorithm of analytical learning which is practically independent from selection of first approximation.

References

[1] Karu, T. (1987), IEEE J. Quentum Electron., QE-23, 1703.

[2] Karu, T. (2003), in Tuan Vo-Dinh (Ed.), *Biomedical Photonics Handbook*, CRC Press, Boca Raton, Ch. 48, 48-1–48-25.

[3] Karu, T. (2008), Photochem. Photobiol. 84, 1091.

[4] Hönigsmann, H. (2009), in Jean Krutmann, Craig A. Elmets, Paul R. Bergstresser. (Eds.), *Dermatological Phototherapy and Photodiagnostic Methods*, Vol. 8, Springer, Berlin.

[5] Hönigsmann, H. (2009). Photo(chemo)therapy for Cutaneous T-Cell Lymphoma, in Hönigsmann, H., Tanew, A., Krutmann J., Hönigsmann H., Craig A., Elmets (Eds.), *Dermatological Phototherapy and Photodiagnostic Methods*, Vol. 8, Springer, Berlin, p. 448, pp. 135–149.

[6] Ortel, B. (2009). "Phototherapeutic Options for Vitiligo" in Ortel, B., Petronic-Rosic, V., Calzavara-Pinton, P. (p. 151), Krutmann J., Hönigsmann H., Craig A., Elmets (Eds.), *Dermatological Phototherapy and Photodiagnostic Methods*, Vol. 8, Springer, Berlin, p. 448, pp. 151–183.

[7] Volkenshteyn, M. V. (1988). *Biophysics*, Second edition, Science, Moskow, p. 591.

[8] Shpolskiy, E. V. (1984). *Atomic Physics*, Science, Moskow, p. 552.

[9] Trunov, A. (1999). "Robust Nonlinear Fault Diagnosis: Application to Robotic Systems, Trunov A., Polycarpou M." *Proceedings of the 1999 IEEE Conference on Control Applications*, pp. 1424–1429.

[10] Trunov, O. M., Belikov, O. E. (2009). "Modeling of interaction EMW with biologics objects in during phototherapy" *Science and Methodology Journal*, – T.107. Vol. 94, (Ecology. – Mykolayiv: Publisher of MSHU named after. Petro Mohyla) pp. 23–27.

[11] Trunov, A. (2012). Patent for invention No. 100029 (Ukraine). Method of Resonance Conformational Photons Therapy and device of its implemantation Appl. 12.05.2010, No. 201005737, publ. 12.11.2012 in Bul. No. 21.

[12] Trunov, A. (2013.). Patent for invention No. 101068 (Ukraine). Method of diagnosis, prevention, rehabilitation and treatment of functions of a tissue when dosed impact magnitude and device for its realization. Appl. 04.05.2011, No. 2011 05583, publ. 25.02.2013, Bul. No 4.

[13] Trunov, A. (2014). Patent for invention No. 104462 (Ukraine). Method of observation of morphology and processes in biological tissues and device of its implemantation, Appl. 10.11.2011, No. 201113252, publ. 10.02.2014, Bul. No. 3.

[14] Fiesler, E. Duong, T., Trunov, A. (2000). Design of neural network-based microship for color segmentation, *IEEE Transaction Intelligent Optical Systems*, Pr. Of SPIE 4055, pp. 228–238.

[15] Nomura, H. Hayashi, I. and Wakami, N. (1992). "A learning method of fuzzy inference rules by descent method," *IEEE*, pp. 203–210.

[16] Alkon, D. L. (1989). Memory Storage and Neural Systems, *Scientific American*, July, 42–50.

[17] Kondratenko, Y. P., Sidenko, Ie. V. (2012). "Correction of the Knowledge Database of Fuzzy Decision Support System with Variable Structure of the Input Data. Modeling and Simulation," Anna M. Gil-Lafuente, V. Krasnoproshin (Eds.)., in Proc. of the Int. Conference MS'12, (2–4 May 2012) Minsk, Belarus (Minsk: Publ. Center of BSU), 56–61.

[18] Kondratenko, Y., Gordienko, E. (2011). "Neural Networks for Adaptive Control System of Caterpillar Turn" Annals of DAAAM for 2011 & Proceeding of the 22th Int. DAAAM Symp. "Intelligent Manufacturing and Automation", (20–23 Oct., 2011, Vienna, Austria), Published by DAAAM International, Vienna, Austria, 305–306.

[19] Kondratenko, Y. P., Sidenko, Ie. V. (2014) "Decision-making based on fuzzy estimation of quality level for cargo delivery", in Recent Developments and New Directions in Soft Computing. Studies in Fuzziness and Soft Computing 317, Zadeh, L. A. et al. (Eds), Springer International Publishing Switzerland, pp. 331–344.

[20] Kryuchkovskiy, V. V., Petrov, K. E. (2011). Development of methodology for identification models of intellectual activity: Problems of information technology, No. 9, pp. 26–33.

[21] Kondratenko Y. P., Encheva S. B., and Sidenko E. V. (2011). "Synthesis of Inelligent Decision Support Systems for Transport Logistic", in Proceeding of the 6th IEEE International Conference on Intelligent Data Acquisition and Advanced Computing Systems: Technology and Applications, IDAACS' 2011, Vol. 2, 15–17 September, Prague, Czech Republic, 642–646.

[22] Atamanyuk I. P., Kondratenko V. Y., Kozlov O. V., and Kondratenko Y. P. (2012) "The algorithm of optimal polynomial extrapolation of random processes," in Modeling and Simulation in Engineering, Economics and Management, K. J. Engemann, A. M. Gil-Lafuente, J. L. Merigo (Eds.), International Conference MS 2012, New Rochelle, NY, USA (30 May–1 June), Proceedings. Lecture Notes in Business Information Processing, Vol. 115, Springer, 78–87.

[23] Trunov, A., Belikov, A. (2015). "Application of Recurrent Approximation to the Synthesis of Neural Network for Control of Processes Phototherapy" in *Proceedings of the IEEE 8th International Conference on Intelligent Data Acquisition and Advanced Computing Systems (IDAACS' 2015)*, Warsow, (September 24–26), Vol. 2, 676–681.

[24] Trounov, A. N. (1990). "Mathematical Aspects of Image Recognition" in *Proceedings Of International Technology* 90, Szezecin, Poland, 479–493.

[25] Rogatkin, D. A., Svirin, V. N., Tchernyi, V. V. (2000). Abstracts Book of 9th International Laser Physics Workshop (LPHYS' 2000), Bordeaux, P. N5.7.

[26] Rogatkin, D. A., Tchernyi, V. V. (2003). Proc. SPIE. Vol. 4955. 554–558.

[27] "A diode array system" http://www.bio-logic.info/assets/brochures/20110421%20-%20DAD%20-%20300dpi.pdf

[28] Trunov, O. M. (2005), "Software constructional features of the express-control systems about the content of substance." *Science and Methtodology Iournal*, ISSN 1609–7742, (Mykolayiv: Publisher of MSHU named after. Petro Mohyla, Ukraine), Vol. 30, 163–176.

[29] Trunov, O. M. (2006), "Software for express-control of ecological safety," in Sience and methtodology journal, ISSN 1609–7742, Trunov O. M (Ed.) Publisher of MSHU named after, Mykolayiv, Petro Mohyla, Ukraine, Vol. 36, 31–37.

[30] Trunov, A. N. (2011) "Recurrent approximation in problems of modeling and design", Monografy – Mykolayiv: Petro Mohyla *BSSU*, 272.

[31] Trunov, A. N. (2016) "Peculiarities of the Interaction of Electromagnetic Waves with Bio Tissue and Tool for Early Diagnosis, Prevention and Treatment", in Proceedings are available in IEEE Xplore Digital Library, IEEE 36th International Conference on Electronics and Nanotechnology (ELNANO) (April), Kyiv, Ukraine, 169–174.

[32] Trunov, A. N. (2013), Intellectualization of the models' transformation process to the recurrent sequence, Eur. Appl. Sci., 1(9), 123–130.

[33] Trunov A. N. (2014), Application of the recurrent approximation method to synthesis of neuron net for determination the hydrodynamic characteristics of underwater vehicles, Prob. Inf. Technol. J., 02(016), 39–47.

[34] Piorek, M. and Winiecky, W. (2015), "On calibration and parametrization of low power ultrawidebanded radar for close range detection of human body and bodity functions", in Piorek, M., Winiecky, W. *Proceedings of the IEEE 8th International Conference on Intelligent Data Acquisition and Advanced Computing Systems (IDAACS'2015)*, Vol. 2, 24–26 September, Warsow, 639–645.

[35] Lyapandra, A. S. Martsenyuk, V. P. Gvozdetska, I. S. and Szklarczyk, R. (2015), "Qualitative analysis of compartmental dynamic system using decision-tree induction", in *Proceedings of the IEEE 8th International Conference on Intelligent Data Acquisition and Advanced Computing Systems (IDAACS'2015)*, Vol. 2, 24–26 September, Warsow, 688–692.

[36] Komar, M., Sachenko, A., Kochan, V. and Skumin, T (2016), "Increasing the resistance of computer systems towards virus attacks", in *Proceedings are available in IEEE Xplore Digital Library, IEEE 36th International Conference on Electronics and Nanotechnology (ELNANO)*, April, Kyiv, Ukraine, 388–390.

[37] Michael W. Kudenov and Eustace L. (2012), "Dereniak compact snapshot real-time imaging spectrometer", in *Proceedings of SPIE*, Vol. 8186 81860W-2, Downloaded from SPIE Digital Library on 06 February to 150.135.50.130. Terms of Use: http://spiedl.org/terms.

[38] Suleimanov, Y. (2016) "Magnetic resonance signal processing tool for diagnostic classification", in Yuri Suleimanov, Sergiy Radchenko,

Oleksandr Lefterov; Andriy Netreba Serhiy Vasnyov Vasyl Sava; Juan Sanchez-Ramos, Leon Prockop, Ranjan Duara (Eds.) *Proceedings are available in IEEE Xplore Digital Library, IEEE 36th International Conference on Electronics and Nanotechnology (ELNANO)*, April, Kyiv, Ukraine, 175–179.

[39] Miroshnichenko, S. I. (2014). Digital Receivers of Rentgen Images, Monograpfy, Publysher "Medicine of Ukraine", Kiev, 98.

[40] Mykola Fisun, Alyona Shved, Yuriy Nezdoliy, Yevhen Davydenko (2015). "The Experience in Application of Information Technologies for Teaching of Disabled Students,", in *Proceedings of the IEEE 8th International Conference on Intelligent Data Acquisition and Advanced Computing Systems (IDAACS'2015)*, Warsow, September 24–26, Vol. 2, 935–939.

[41] Dzyuba, D. A., Chernodub, A. N. (2011). Application of methods active initiations for modification neuro-controler in real time, Mathemation's mashine and system, Kiev, No. 1, 20–28.

[42] Bodyanskiy, Ye., Chaplanov, O., Popov, S. (2003). "Adaptive prediction of quasiharmonic sequences using feedforward network" in *Proceedings of the International Conference of Artificial Neural Networks and Neural Information Processing ICANN, (ICONIP 2003)* Istanbul, 378–381.

[43] William H. Smith (1990). Digital array scanned interferometer US 4976542 A, published 11 December.

Index

About the Editors

Piotr Bilski (Ph.D., D.Sc.) was born in 1977 in Olsztyn, Poland. He graduated from Warsaw University of Technology, Institute of Radioelectronics, obtaining M.Sc. degree in 2001 (with honors), Ph.D. degree in 2006 (with honors) and D.Sc. degree in 2014 in computer science. Currently he is an Associate Professor in the Institute of Radioelectronics and Multimedia Technologies, Warsaw University of Technology. His scientific interests include diagnostics of analog systems, design and analysis of virtual instrumentation, application of artificial intelligence and machine learning methods to the acoustics and environmental sciences. He is the member of IEEE, IMEKO TC10 and POLSPAR and reviewer for such journals like Measurement, IEEE Transactions on Instrumentation and Measurement, and Expert Systems with Applications.

Francesca Guerriero is full professor of Operations Research at University of Calabria, Italy. Her research interests include the application of mathematical programming to problems arising in logistics, healthcare systems, and telecommunication and transportation networks. She has published over 90 manuscripts, many of which appeared in top-tier journals.

About the Authors

Agata Kubik received her M.Sc. Eng. degree in medical physics at the Faculty of Physics at the Warsaw University of Technology in 2015. She is currently a Ph.D. student at the Institute of Radioelectronics and Multimedia Technology at Warsaw University of Technology. Her research focuses on development of new analysis methods in 4d flow cardiac magnetic resonance imaging.

Aleksander Volkov was born in 1987 in Novolavela village of Pinezhsky region, Russia. He graduated from Pomor State University named after M. V. Lomonosov as a Physics and Informatics teacher in 2010, in 2015 he obtained Master's degree of Biology (educational program "Psycho-physiology") having defended the thesis "CFFF as a method of psycho-physiological studying of a visual analyzer". Currently he holds a position of the head of General and Applied Physics Department laboratories of Higher School of Natural Sciences and Technologies, Northern (Arctic) Federal University named after M. V. Lomonosov. Scientific interests include Physics of dispersion medium, psycho-physiological peculiarities of visual perception, colour vision, age-related physiology, human adaptation to circumpolar region conditions. Volkov A. is a participant of the pilot project "Arctic floating university – 2012" on the research ship "Professor Molchanov".

Alexander Trunov, Ph.D., associate professor, engineer. Lecturer, associate professor of physics, head of department "Means of adaptability and sensors" of division "Deep underwater technology" MSI named after Admiral S. Makarov (1975–1997 years), supervisor of projects under the SCST USSR "Durability and reliability mashine and mechanisms", "The world ocean development", "Robotics", including "Uranium-1".

Dean of the Faculty of Computer Science NaUKMA (1997–1999 years.).

Co-Head of US-Ukrainian project on analysis of latest technology materials, with the support of the US Civilian Research and automated systems, including early diagnosis and physiotherapy.

From 2000 to 2007 years – Vice-Rector and from 2007 till now the First Vice-Rector, associate professor of medical devices and systems of Petro Mohyla Black Sea Natonal University, author of over hundred fifty scientific papers and manuals, including monography and book, 29 patents, including patents and Ukraine.

Scientific interest: Nonlinear problems; nonlinear mathematical programming; robotics, sensor, automation, recurrent approximation, recurrent neuron net.

Alexey Lagunov (Candidate of Pedagogic Sciences) was born in 1961 in Arkhangelsk, Russia. He graduated from Arkhangelsk State Pedagogic Institute, Pomor State University named after M. V. Lomonosov as a mathematics teacher in 1988, obtained candidate of pedagogic sciences in 1997 having defended the thesis "Methodology of electronic tables usage for school mathematical problems solution" (Moscow). Since 2000 he is an Associate Professor and currently the head of Microsystem Technics and Digital Technologies Department of Higher School of Natural Sciences and Technologies, Northern (Arctic) Federal University named after M. V. Lomonosov (Arkhangelsk, Russia; abbreviated form: "NArFU named after M. V. Lomonosov" or "NArFU"). His scientific interests include programming, physical basics of microsystem technics, medical informatics, alternative energy, adaptation of technical devices to circumpolar region conditions. He is a chairman of the regional expert group of Informatics education in Arkhangelsk region.

Angelo Consoli received M.Sc. in electronic engineering from ETHZ Zürich. Professor and a Head of the IT security laboratory at the University of Applied Science of Southern Switzerland. Industrial experience: Siemens AG (4.5 years), AIG (3 years) and many years as freelance technology consultant. Aerospace Programme Manager, project leader and head of the space business unit for NemeriX (2001–2009). Founder and managing director of ECLEXYS Sagl. Since 2007, mandated by the Swiss Government as a member of the group of experts entitled by the Swiss Space Office for related projects.

Dmitry Fedin (engineer) was born in 1989 in Arkhangelsk, Russia. He graduated from Northern (Arctic) Federal University named after M. V. Lomonosov in 2013 as an engineer of Information systems and technologies and in 2015 he obtained Master's degree of Physics, (educational program "Information processes and systems"), Northern (Arctic) Federal University named after M. V. Lomonosov having defended the thesis "Telemetry data collecting complex development for satellite signal coverage area monitoring". Currently he holds a position of the engineer of Radiotechnical Monitoring Centre of Higher School of Natural Sciences and Technologies, NARFU named after M. V. Lomonosov. His scientific interests include programming, alternative energy, technical device adaptation to circumpolar region conditions. Fedin D. is a participant of the pilot project "Arctic floating university – 2013" on the research ship "Professor Molchanov".

Filomena Olivito earned a Ph.D. in Operations Research at University of Calabria, Italy. She was also involved in tutoring and project activities at the Department of Mechanical, Energy and Management Engineering at the same University.

Francesco Piazza received the M.Sc. degree in Electrical Engineering from the Swiss Federal Institute of Technology (ETHZ), Zurich, and the Ph.D. degree from the Integrated Systems Laboratory at the same university in 1999 under the supervision of Prof. Qiuting Huang. From 2000 to 2002 he worked for TChip Semiconductor as RF/analogue IC designer and from 2002 to 2008 at NemeriX SA as a chief scientist for RF design. Since 2009 he is with Saphyrion Sagl as a chief scientist, where he works in the development and industrialization of rad-hard space qualified RF/analog integrated circuits.

Frode F. Jacobsen, Ph.D., RN, is an anthropologist and Professor in older people's care at Western Norway University of Applied Sciences and Professor II at VID Specialized University, Norway. Jacobsen is Research Director of the Center for Care Research – Western Norway. Jacobsen's research interests are health and health seeking behavior in a social and cultural context, comparative health services systems, and, health and care services for older people.

Giovanna Miglionico is a research assistant at the Logistics Laboratory of the University of Calabria, Italy. Her research interests include nonlinear and nonsmooth optimization, revenue management, and optimization models for decision making in logistics.

Ingebjørg T. Børsheim is an experienced occupational therapist with a master in social work. She is Assistant professor at Western Norway University of Applied Sciences, Department of Occupational Therapy, Physiotherapy and Radiography. Her research interests are related to how people's daily activity are of importance for health, well-being and participation and how technology can provide assistance for people with impaired function to cope with everyday demands.

Jakub Wagner (Student Member, IEEE) received the M.Eng. degree in biomedical engineering from the Warsaw University of Technology, Poland, in 2013. He is currently pursuing the Ph.D. degree at the Institute of Radioelectronics and Multimedia Technology, Warsaw University of Technology. His principal research interests include measurement data processing in medical and healthcare applications.

Jan Jakub Szczyrek, M.Sc., graduated at Warsaw University of Technology (Electronics in 1994) and Warsaw University (Mathematics in 1997). Main fields of expertise: applied cryptography, dynamical systems – especially quasiconformal dynamics, real time operating systems on embedded platforms and wireless telecommunication. Activities: chief of development team in Sylan project (encrypting telecommunication system including different media and Public Key Infrastructure), participation in AURA project (radar system dedicated to UAVs) and RADCARE system.

Jaouhar Ayadi received the Engineer Degree ('94) in electrical engineering from the Ecole Nationale d'Ingénieurs de Tunis in 1994. He received the M.Sc. ('95) and Ph.D. ('99) in Computer Science and Communication from the Ecole Nationale Supérieure des Télécommunications (ENST)-René Descartes University, Paris. In 1999, he joined the Wireless Communication Section of the Centre Suisse d'Electronique et de Micro-technique (CSEM), Switzerland, Currently he is a R&D Manager at ECLEXYS. His research interests include Digital Signal Processing for mobile communications, wireless propagation, UWB systems.

Jerzy Kołakowski received the M.Sc. ('88) and Ph.D. ('00) degrees in telecommunications from the Warsaw University of Technology. Since 1988 he has been with the Institute of Radioelectronics and Multimedia Technology where he holds a position of Assistant Professor. He is a Member of the Management Board of the Foundation for the Development of Radiocommunications and Multimedia Technology, his current research interests include positioning systems, ultra wideband technology, cellular systems.

Karol Radecki received the M.Sc. and Ph.D. degrees in electronic engineering from the Warsaw University of Technology, Poland, in 1970 and 1977, respectively. Since 1970 he has been with the Warsaw University of Technology, where he is an Assistant Professor at the Institute of Radioelectronics and

Multimedia Technology. In the years 1995–2011 he was the URSI Commission A (Electromagnetic Metrology) National Chairman. His field of interest covers atomic frequency standards and electronic orientation systems for elderly and blind people.

Professor Knut Øvsthus, Ph.D., has a background as scientist at Telenor R&D and FFI, from 2006 he has hold the position as Professor at Western Norway University of Applied Sciences. His Ph.D. research was in semiconductor laser, conduction both theoretical and experimental research. Following this research he has conducted research in Internet technologies, both wireless ad-hoc network and wireless sensor networks. He has had several research projects and collaborated in international research activities.

Konrad Werys received his Ph.D. in electronic engineering at the Faculty of Electronics and Information Technology at the Warsaw University of Technology in 2016. His research focuses on developing new quantification measures in cardiac magnetic resonance imaging. Clinical applications of the new methods are tested in cooperation with the Cardinal Stefan Wyszyński Institute of Cardiology in Warsaw.

Lorenzo Moriggia received the B.S. degree in Electrical Engineering from the University of Applied Sciences and Arts of Southern Switzerland (SUPSI), Manno in 2004. From 2004 to 2009 he worked as Researcher at the Department of Innovative Technologies of SUPSI. Since 2009 he is with Saphyrion Sagl as a principal HW electronic engineer, where he works in the development of RF and digital boards.

Ludmila Morozova (Doctor of Biological Sciences, Professor) was born in 1968 in Novodvinsk, Russia. She graduated from Arkhangelsk State Pedagogic Institute, Pomor State University named after M. V. Lomonosov as a biology and chemistry teacher in 1990, obtained Candidate of Biological Sciences in 1995 and Doctor of Biological Sciences in 2008. The theme of a thesis defence is "Psycho-physiological regularities of visual perception of 6–8 years old children" (Arkhangelsk). Currently she is the Professor of Higher School of Natural Sciences and Technologies, Northern (Arctic) Federal University named after M. V. Lomonosov (Arkhangelsk, Russia). Scientific interests include psycho-physiological peculiarities of visual perception, colour vision, age-related psycho-physiology, human adaptation to circumpolar region conditions. Morozova L. is a member of Physiology Community named after I. P. Pavlov, member of Inter-regional Association of Cognitive Researcher, expert of the Russian Academy of Sciences.

Łukasz Błaszczyk received his M.Sc. Eng. degree in biomedical engineering at the Faculty of Electronics and Information Technology at the Warsaw University of Technology in 2013. He is currently an Assistant at Faculty of Mathematics and Information Science and a Ph.D. student at the Institute of Radioelectronics and Multimedia Technology at the Warsaw University of Technology. His research focuses on compressed sensing, applications of hypercomplex algebras in signal processing and applied mathematics.

Lukasz Malicki a vice president of Knowledge Society Association where works as a work package leader and product manager in several R&D projects (e.g. NITICS AAL-JP, CAMI AAL-JP). In SSW he is responsible for end-user requirements and e-health service concepts, designing key functionalities and working as an expert in user requirement gathering. Previously he worked as an academic researcher and teacher and was an owner of an innovation microenterprise. His approach comes mainly from a background in biomedical engineering and management. Moreover, he has been developing the new paradigm of designing applications and websites that are meant to be used by all end-users (especially seniors, public administration, hospitals), which will result in an increased accessibility for people who are totally or partially deprived of access to digital media.

Magdalena Berezowska was born in Poland in 1990. She received B.Sc. degree in Electrical and Computer Engineering and M.Sc. degree in Telecommunications (with honors) from Warsaw University of Technology, Poland in 2013 and 2015, respectively. Since then, she has been focusing on the area of IoT solutions and low-power data exchange protocols. Her main areas of interest include biometric sensors, human-sensor interactions and low-power wearable sensor applications.

Nadejda Podorojnyak (engineer) was born in 1990 in Arkhangelsk, Russia. She graduated from Northern (Arctic) Federal University named after M. V. Lomonosov in 2012 as an engineer of Information systems and technologies, in 2012 as a "Translator in the field of professional communication" and in 2014 she obtained Master's degree of Management Science (educational program "State and municipal administration") in Northern (Arctic) Federal University named after M. V. Lomonosov having defended the thesis

"E-government's work improvement on a regional level". Currently she holds a position of the engineer of Radiotechnical Monitoring Centre of Higher School of Natural Sciences and Technologies, Northern (Arctic) Federal University named after M. V. Lomonosov. Her scientific interests include translation of scientific papers, work process optimization problem solving, medical informatics, alternative energy, adaptation of technical devices to circumpolar region conditions.

Paweł Mazurek received the B.Sc. degree in data communications and telecommunication management, in 2012, and the M.Sc. degree in telecommunications, in 2014, both from the Faculty of Electronics and Information Technology, Warsaw University of Technology, Warsaw, Poland, where he is working toward the Ph.D. degree in electronics. His research interests include the preprocessing of measurement data acquired by means of impulse radar sensors and infrared depth sensors and the identification of the potential behind the fusion of those sensors when applied for patients monitoring.

Piotr Bogorodzki is an associate professor at the Faculty of Electronics and Information Technology at the Warsaw University of Technology. His research focuses on software and hardware developments in magnetic resonance imaging. These covers either high frequency designs in hyperpolarization or processing methods in computational neuroanatomy. He is also responsible for MRI data processing in several grants in neuroimaging area.

Ryszard Michnowski received the M.Sc. degree and Ph.D. degree in electronics and electrical engineering from the Warsaw University of Technology, Warsaw, Poland, in 1997 and 2006, respectively. In 2007, he joined the AM Technology as a sales advisor. From 2008 he has been an Assistant Professor with the Institute of Radioelectronics and Multimedia Technology, Warsaw University of Technology. His research interests include microwave and millimeter-wave circuits and UWB technologies. He has coauthored more than 30 conference scientific papers.

Stanisław Jankowski, Ph.D., D.Sc., professor at the Warsaw University of Technology, Faculty of Electronics and Information Technology, Warsaw, Poland. Expert in: artificial intelligence, head of the research group on statistical learning systems. Area of research: theory – complex-valued associative memory, cellular neural networks, kernel machines for regression and classification, recurrent neural networks and recurrent least-squares support vector machines, optimisation of learning systems based on influential statistics, memristive neural networks, semi-supervised learning; applications – computer-aided medical diagnosis, identification of defects in semi-isolating electronic materials, robotics and aviation, geology, seismology, healthcare systems.

Tobba T. Sudmann is Associate Professor, physiotherapist and sociologist, with a Ph.D. from Department of Global Health and Primary Care at the University of Bergen. She is currently affiliated to Department of Social Pedagogy and Social Work at Western Norway University of Applied Sciences. Her research interest are related to how people use their bodily resources to enhance their well-being and social participation, whether the means are physical activity, technology or animals. Publications and research are directed towards the person(s)'s agency, and towards anti-oppressive professional practice.

Tomasz Ciamulski is a research engineer working on applications of electronic technologies to industrial as well as consumer problems. He created and executed several international R&D projects involving industrial and academic cooperation. Results of electromagnetic research he applied commercially in WiSub AS company as its technology co-founder, creating an innovative solution for pinless underwater connectors. He also seeks for applications of ambient sensors and other assisted technologies for diagnostic and life quality improvement of elderly while supporting cross-disciplinary cooperation between Warsaw University of Technology (Ph.D. origin) and Western Norway University of Applied Science.

Uladzimir Dziomin is a lecturer of Intelligent Information Technology Department of Brest State Technical University. He received his M.Sc. and Diploma in the same department. He is working on Ph.D. thesis in Neural Networks, Reinforcement Learning and Multi-Agent systems at laboratory of Robotics in BrSTU. His research interests include Mobile Robotics, Distributed systems and Swarm Intelligence.

Vitomir Djaja-Josko received the B.Sc. ('13) and M.Sc. ('15) degrees in telecommunications from the Faculty of Electronics and Multimedia Technology, Warsaw University of Technology, Poland. He is pursuing the Ph.D. degree in the Institute of Radioelectronics and Multimedia Technology. He is currently a member of the research team working in the field of UWB technologies. His research interests include UWB signals, indoor localization and synchronization methods in positioning systems.

Vladimir Terehin (engineer) was born in 1989 in Arkhangelsk, Russia. He graduated from Northern (Arctic) Federal University named after M. V. Lomonosov in 2015 as an engineer of Information systems and technologies, at the present time he studies for Master's degree in Physics, educational program – "Biophysics and medical devices". Currently he holds a position of the engineer of Radiotechnical monitoring centre of Institute of Natural Sciences and Technologies, Northern (Arctic) Federal University named after M. V. Lomonosov. His scientific interests include programming, electrical engineering, physical basics of microsystem technics, medical informatics, biophysics, alternative energy, technical device adaptation to circumpolar region conditions. Terehin V. is a participant of the pilot project "Arctic floating university – 2013" on the research ship "Professor Molchanov".

Prof. Wiesław Winiecki, M.Sc. (1975), Ph.D. (1986), D.Sc. (2003); Prof. Title (2011), Professor of Measurement Science at the Institute of Radioelectronics, Warsaw University of Technology, has fourty-year research experience in the field of measurement and control systems, including the development and implementation of various kinds of measurement devices and systems. The record of his achievements in this respect comprises more than 200 research publications, 2 monographs, 1 book and 1 academic textbook, as well as over 100 reports on scientific research and implementation. He is the Vice-President of the Polish Society for Measurement, Automatic Control and Robotics POLSPAR.

In the years 1994–2001 and 2004–2005, he held the position of Deputy Director for Research at the Institute of Radioelectronics at Warsaw University of Technology (WUT), then, in 2005–2008 he was Vice-Dean for Research in the Faculty of Electronics and Information Technologies, WUT. In 2008, he has been re-appointed as Deputy Director for Research at the Institute of Radioelectronics. From 2016 he holds the position of Director at the Institute of Radioelectronics and Multimedia Technology at WUT.

Zbigniew Szymański, M.Sc., senior lecturer at the Institute of Computer Science, Faculty of Electronics and Information Technology at Warsaw University of Technology. Research interests include: computational intelligence, kernel machines for regression and classification, recurrent neural networks and recurrent least-squares support vector machines, optimisation of learning systems based on influential statistics and embedded systems. He participated in research projects in the field of computer-aided medical diagnosis, aviation, geology, seismology, where computational intelligence was applied.

CPSIA information can be obtained
at www.ICGtesting.com
Printed in the USA
BVOW07*2245280417
481668BV00004B/2/P